# Practical Laboratory Mycology

**Third Edition**

# Practical Laboratory Mycology

**Third Edition**

**ELMER W. KONEMAN, M.D.**

*Associate Professor of Pathology*
*Northwestern University School of Medicine*
*Chicago, Illinois*

**GLENN D. ROBERTS, Ph.D.**

*Associate Professor of Laboratory Medicine and Microbiology*
*Consultant, Section of Clinical Microbiology*
*Mayo Medical School*
*Mayo Clinic and Mayo Foundation*
*Rochester, Minnesota*

SANS TACHE
WILLIAMS & WILKINS
Baltimore • London • Los Angeles • Sydney

*Editor:* George Stamathis
*Associate Editor:* Carol Eckhart
*Copy Editor:* Shelley Potler
*Design:* Bert Smith
*Illustration Planning:* Reginald R. Stanley
*Production:* Anne G. Seitz



Accurate indications, adverse reactions, and dosage schedules for drugs are provided in this book, but it is possible that they may change. The reader is urged to review the package information data of the manufacturers of the medications mentioned.

*Made in the United States of America*

Second Edition, 1978
Reprinted 1979, 1981, 1982, 1983

**Library of Congress Cataloging in Publication Data**

Koneman, Elmer W., 1932-
Practical laboratory mycology.

Bibliography: p.
Includes index.
1. Medical mycology—Laboratory manuals. 2. Fungi, Pathogenic. I. Roberts, Glenn D. II. Title. [DNLM: 1. Mycology—laboratory manuals. QW 25 K815p]
QR245.K66 1985 616′.015′028 84-19517
ISBN 0-683-04746-9

Composed and printed at the
Waverly Press, Inc. 85 86 87 88 89 10 9 8 7 6 5 4 3 2 1

# PREFACE TO THE THIRD EDITION

The intent of the Third Edition of *Practical Laboratory Mycology* is to improve on the Second Edition, incorporating current concepts and format changes so that the material can be utilized more effectively by beginning students in clinical mycology. Retained from the First Edition is the practical approach of dividing the fungi of medical importance into several subgroups based on certain colonial and microscopic characteristics, after which genus identification can be made by comparing the unique morphologic features of each member in the group.

To this end, the basic organization of the previous editions has been retained in the Third Edition. The functional step-by-step approach to establishing an identification of a fungal isolate and providing the basis for establishing a clinical diagnosis is the same as in the Second Edition. The chapters again follow in sequence discussions of the clinical signs and symptoms that lead a physician to suspect a mycotic infection, specimen collection and processing, media selection, methods for inoculation and incubation of cultures, methods for preparing mounts for microscopic study, and a description of the morphologic characteristics of the medically important fungi. Chapter 3, introducing the direct examination of clinical specimens, has been added in this edition to help medical technologists and microbiologists more ably assist pathologists in making the identification of fungal elements in unstained and stained preparations.

The line drawings of the microscopic features of the fungi discussed in the text have been replaced by photomicrographs which better illustrate the morphologic characteristics as they are actually seen under the microscope. Most of the photographs used in the Third Edition, including the photomicrographs, are grouped in several plates so that common subject matter can be compared. For example, placing the photomicrographs of the groups and subgroups of fungi outlined in the practical working schema included in Chapter 5 in juxtaposition adjacent to one another in these plates makes it convenient for the reader to compare the subtle microscopic differences that separate the various members within each functional group. A brief summary of the key diagnostic features follows the discussion of each fungus included in this text, a format change that should aid students in the quick recall of the essential information necessary to make their identification.

Appendix I, the "Glossary of Useful Mycological Terms," Appendix II, a listing of the formulae for the various media, reagents and stains used in the study of fungi, and the Color Plates of representative fungal colonies have been retained from the Second Edition with minor editing. The various diagnostic procedures are incorporated in the running text as they were in the Second Edition. Dropped from the Second Edition is the chapter on the filamentous fungus-like bacteria. The *Actinomycetes,* including *Nocardia* sp and *Streptomyces* sp belong to the bacterial class *Schizomycetes.* It is our feeling that to include a discussion of these microorganisms in a mycology text is not appropriate.

The authors have adopted most of the newer names and terms as reflected in the recent mycological literature. *Acremonium, Drechslera, Pseudallescheria, Candida glabrata, Wangiella,* and *Exophiala* are among the new genus designations adopted in this text. Where appropriate, the previous designation is placed in parentheses to that students will not be confused when recalling the old name from the literature. The terms phialides and metulae are used instead of sterigmata to describe hyphal segments supporting conidiogenesis; however, we have reverted back to the use of microconidia and macroconidia in deference to the terms microaleuriospore and macroalleuriospore.

As was the case with the previous editions, the Third Edition also is designed as an introductory text in laboratory mycology, serving to provide new mycologists with the fundamental knowledge needed to perform in a clinical microbiology laboratory and as a prerequisite course of study for a better understanding of more advanced textbooks in mycology. The authors have incorporated many of the positive suggestions and constructive criticisms that have been received relative to the previous editions. We also wish to thank our laboratory staffs and workshop participants who have found our approach to the study of mycology helpful both during their formative period of instruction and as a practical guide in their practice and teaching activities.

The external force that will influence clinical laboratory practice in the decade ahead is cost containment and the need for laboratory directors and supervisors to select carefully only those procedures necessary to establish the identifications of organisms recovered in clinical cultures. We believe that this text innately provides that focus. On page viii is a listing of 95 species of fungi that have been recovered from respiratory specimens at the Mayo Clinic during the period 1975 to 1980 in decreasing order of frequency. This experience is not unlike that of others and reflects what might be encountered in clinical laboratories where a concerted effort is made to recover these etiologic agents. This list also provides a valuable guideline to laboratory directors who wish to know where to apply their laboratory resources and to instructors who must plan the curricula to be used in mycology training programs.

# PREFACE TO THE SECOND EDITION

The Second Edition of *Practical Laboratory Mycology* serves not only to update certain sections of the First Edition, but also provides a more effective stand alone text for self-study and student teaching. The First Edition was designed as a supplement to the Medcom Famous Teachings in Modern Medicine slide series, *Clinical Laboratory Mycology, Parts I and II,* and was dependent in large degree on the availability of the color slides for maximum learning.

The First Edition sections on specimen collection, selection and use of media, culture processing, and identification of yeasts have been largely rewritten. Over 80 photomicrographs to supplement the many line drawings of microscopic morphology and 5 color plates including 40 color prints of frequently encountered fungal colonies have been added. A supplemental appendix listing the formulas for commonly used media, stains, and reagents should also prove of value.

The basic organization of the first Edition has been retained in the Second Edition. The logical step by step discussion of specimen collection and processing, media selection, methods for inoculation and incubation of cultures, and the practical approach to the identification of fungi as presented in the Second Edition follows the natural flow of work in clinical mycology laboratories. Thus, instructors of medical mycology students are not only provided with a practical curriculum guide, but one that follows closely the future work patterns in the laboratory.

The authors of this Second Edition continue to find in their teaching programs and workshop sessions that the subgrouping of the 75 to 100 species of fungi of medical importance based on readily observably gross culture and microscopic characteristics materially aids in making the initial study of mycology less complex for new students.

The design and content of the manual remain as an introductory text in laboratory mycology. It should serve well as a prerequisite course of study for a better understanding of more advanced textbooks in mycology, or as a beginning for interested students and laboratory technologists to pursue a career in mycology or engage in research and advanced studies.

## *Fungi Isolated from Respiratory Tract Specimens (Mayo Clinic 1975 to 1980)*

| ORGANISM | NO. ISOLATES | ORGANISM | NO. ISOLATES |
|---|---|---|---|
| 1. Yeast not *Cryptococcus* | 19204 | 48. *Fusidium* spp.* | 9 |
| 2. *Penicillium* spp. | 5102 | 49. *Ostracoderma* spp.* | 9 |
| 3. *Cladosporium* spp.* | 3208 | 50. *Oidiodendron* spp.* | 8 |
| 4. *Aspergillus* spp. | 2340 | 51. *Nigrospora* spp.* | 7 |
| 5. *Aspergillus fumigatus* | 2223 | 52. *Scopulariopsis brumptii* | 7 |
| 6. *Alternaria* spp.* | 1503 | 53. *Sporothrix* spp.* | 7 |
| 7. *Aspergillus niger* | 896 | 54. *Torulomyces* spp.* | 7 |
| 8. *Fusarium* spp. | 689 | 55. *Circinella* spp.* | 5 |
| 9. *Geotrichum* spp. | 577 | 56. *Gliomastix* spp.* | 5 |
| 10. *Aspergillus versicolor* | 559 | 57. *Prototheca* spp. | 5 |
| 11. *Beauveria* spp.* | 457 | 58. *Acremoniella* spp.* | 4 |
| 12. *Scopulariopsis* spp. | 444 | 59. *Cryptococcus terreus* | 4 |
| 13. *Aspergillus flavus* | 442 | 60. *Paracoccidioides brasiliensis* | 4 |
| 14. *Acremonium* spp.* | 429 | 61. *Exophiala jeanselmei** | 4 |
| 15. *Mucor* spp. | 377 | 62. *Sporothrix schenckii* | 4 |
| 16. *Paecilomyces* spp.* | 302 | 63. *Dematium* spp.* | 3 |
| 17. *Rhizopus* spp. | 265 | 64. *Stachybotrys* spp.* | 3 |
| 18. *Cryptococcus neoformans* | 205 | 65. *Absidia* spp. | 2 |
| 19. *Nocardia asteroides*† | 157 | 66. *Acrostalagmus* spp.* | 2 |
| 20. *Cryptococcus albidus* var. *albidus* | 145 | 67. *Aspergillus candidus* | 2 |
| 21. *Streptomyces* spp.† | 139 | 68. *Botrytis* spp.* | 2 |
| 22. *Histoplasma capsulatum* | 137 | 69. *Cunninghamella* spp.* | 2 |
| 23. *Coccidioides immitis* | 126 | 70. *Graphium* spp.* | 2 |
| 24. *Trichoderma* spp.* | 109 | 71. *Helminthosporium* spp.* | 2 |
| 25. *Blastomyces dermatitidis* | 84 | 72. *Hanseniaspora* spp.* | 2 |
| 26. *Aspergillus terreus* | 77 | 73. *Malbranchia* spp.* | 2 |
| 27. *Aspergillus glaucus* | 75 | 74. *Monascus* spp.* | 2 |
| 28. Unidentified aerobic non-acid fast actinomycete† | 73 | 75. *Monascus purpureus** | 2 |
| | | 76. *Nocardia brasiliensis*† | 2 |
| 29. *Phoma* spp.* | 71 | 77. *Sepedonium* spp.* | 2 |
| 30. *Syncephalastrum* spp.* | 58 | 78. *Tritirachium* spp.* | 2 |
| 31. *Aspergillus clavatus* | 52 | 79. *Abgliophragma* spp.* | 1 |
| 32. *Trichosporon* spp. | 45 | 80. *Actinomucor* spp.* | 1 |
| 33. *Pseudallescheria boydii* | 42 | 81. *Bispora* spp.* | 1 |
| 34. *Papularia* spp.* | 38 | 82. *Gliocladium* spp.* | 1 |
| 35. *Aureobasidium pullulans** | 35 | 83. *Gymnoascus* spp.* | 1 |
| 36. *Chrysosporium* spp.* | 35 | 84. *Heterosporium* spp.* | 1 |
| 37. *Aspergillus nidulans* | 33 | 85. *Monilia* spp.* | 1 |
| 38. *Epicoccum* spp.* | 33 | 86. *Monocillium* spp.* | 1 |
| 39. *Rhinocladiella* spp.* | 32 | 87. *Neurospora* spp.* | 1 |
| 40. *Stemphylium* spp.* | 31 | 88. *Nocardia* spp. | 1 |
| 41. *Phialophora* spp.* | 27 | 89. *Phyctaenia* spp.* | 1 |
| 42. *Botryoderma* spp.* | 22 | 90. *Pithomyces* spp.* | 1 |
| 43. *Curcularia* spp.* | 19 | 91. *Scolecobasidium* spp.* | 1 |
| 44. *Drechslera* spp. | 16 | 92. *Sporobolomyces* spp.* | 1 |
| 45. *Stysanus* spp.* | 12 | 93. *Trichothecium* spp.* | 1 |
| 46. *Verticillium* spp.* | 11 | 94. *Trichothecium roseum** | 1 |
| 47. *Wangiella dermatitidis** | 10 | 95. *Trichurus spiralis** | 1 |

* Filamentous fungi which are not recognized respiratory tract pathogens.
† Filamentous bacteria belonging to the class Schizomyces.

# ACKNOWLEDGMENTS

The authors wish to acknowledge the dedication and extraordinary work performed by Leslie Stockman, Research Technologist, Section of Clinical Microbiology at the Mayo Clinic for preparing the sections and mounts for photomicrography, for selecting and cropping of the final illustrations and for proof reading of the text.

To whatever extent the color plates in this text live into perpetuity, the name of Benjamin F. Summers, who was the photographer at the Presbyterian Medical Center, Denver, Colorado at the time many of the original color prints were made, should be placed in memorium in commemoration of his death earlier this year.

Also, our appreciation is extended to the American Society of Clinical Pathologists for granting permission to reproduce many of the color prints found in the Color Plates, originally published in *Atlas of Clinical Mycology*, by C. T. Dolan, J. W. Funkhouser, E. W. Koneman, N. G. Miller, and G. D. Roberts, American Society of Clinical Pathologists, Chicago, 1975.

# CONTENTS

# Chapter 1

# DIAGNOSIS OF MYCOTIC DISEASE

## INTRODUCTION

We have suggested three major areas of activity that must be coordinated to achieve the highest probability of establishing the diagnosis of a mycotic disease (44):

1. **The clinical setting** is a situation where the patient presents with signs and symptoms suspicious of mycotic disease, such as weight loss, lassitude, cough, localized pain or lesions of the skin and/or mucous membranes. In turn, the physician takes a medical history, performs a physical examination, orders X-rays and laboratory tests and obtains appropriate specimens for culture.
2. **The surgical pathology laboratory** is where a search for fungal elements is made in cytological preparations of fluid specimens, in frozen sections of biopsies or in stained sections prepared from resected tissues and organs. A portion of tissues or organs suspected of one of the mycoses should be submitted for fungal culture.
3. **The microbiology laboratory** is where specimens are examined and cultures set up to recover fungi. Growth appearing on culture media is examined grossly and mounts are prepared for microscopic study to derive a final identification. Biochemical characteristics may be assessed to confirm the differential identification of some species of fungi.

### The Clinical Setting

Because fungal spores or hyphal fragments easily become airborne and can be readily inhaled, the lungs are the initial site of infection for most mycoses and patients often present with pulmonary signs and symptoms. Chronic cough, with or without sputum production, chest pain, and dyspnea are common symptoms. The persistence of a pulmonary infiltrate on X-ray, or the presence of a cavity containing a "fungus ball" or the appearance of a "coin lesion" may represent either an inactive, latent, or slowly progressive infection that may require further investigation.

Wheezing, asthma-like attacks or expectoration of thick mucous plugs are manifestations of allergic bronchopulmonary mycosis. Such symptoms may not indicate an endogenous fungal infection, rather they may be caused by inhalation of dust containing spores or mycelial elements to which a hypersensitive host reacts.

The skin is a common site of fungal disease caused primarily by a group of fungi called **dermatophytes** that produce dermatoses or what are also called "**ringworm**" infections. Most individuals have experienced the itching, scaling, weeping manifestations of athlete's foot, may have had "jock itch," or the circular, red ring-like lesions typical of classic ringworm of the skin. Focal loss of hair in cases of tinea capitis or fungal involvement of the nails are less common manifestations.

Deeper skin or subcutaneous fungal infections are less commonly encountered. However, mycotic disease must be suspected with any progressive, nonhealing ulcer of the skin or mucous membranes, or when deeply penetrating sinuses exuding a purulent exudate are encountered. One should particularly be suspicious of fungal infections if post-traumatic subcutaneous or ulcerative wounds are contaminated with soil or vegetative matter. Skin penetration with thistles, thorns, burrs, or similar materials creates a particular risk for the development of cutaneous or subcutaneous mycotic infections. The physician must remember that these cutaneous lesions may not represent primary disease, but rather may be an extension of a serious disseminated fungal infection. When deep cutaneous lesions suspicious of mycotic disease are encountered, a careful physical examination of the patient must be performed, chest X-rays obtained, and specimens procured for culture, not only from the skin lesions but also of sputum, urine, blood, and other infected body sites.

Certain skin lesions, such as the painful erythematous areas seen in **erythema nodosum** or the symmetrical macular, papular, or vesicular eruptions of the extremities in **erythema multiforme**, should alert one to the possibility of fungal infections. Symmetrical vesicular skin lesions of the hands or feet—a clinical manifestation called the **"id" reaction**, a term derived from the word "dermatophytid"—represent a cutaneous hypersensitivity reaction to bacterial or fungal infections that may involve other parts of the skin or deep viscera. The physician must, therefore, perform a physical examination, order X-rays, and procure cultures of other infected body sites when an id reaction is seen. The id lesions do not contain fungi and produce negative cultures.

General systemic signs and symptoms may suggest the presence of mycotic disease. Fever, night sweats, weight loss, lassitude, and easy fatigability are common general symptoms. Tuberculosis, syphilis, sarcoidosis, and carcinomatosis may have similar clinical manifestations and must be considered in the differential diagnosis. Low grade persistent peripheral blood leukocytosis, particularly eosinophilia and monocytosis, an increased sedimentation rate, or increase in one or more of the serum gamma globulin fractions all suggest the possibility of mycotic infection.

In summary, all immunosuppressed patients with fever and signs of systemic infection should be evaluated for the presence of fungal infection.

### The Surgical Pathology Laboratory

Descriptions of the histopathology of fungal infections will be included in later chapters of this text. Because the study of stained tissue sections is an activity generally considered outside of the realm of the clinical mycology laboratory, a discussion of how fungi appear in biopsy specimens was omitted from the previous editions of this text. It has been brought to our attention, however, that medical technologists and clinical microbiologists are often asked by surgical pathologists to examine tissue sections and express an opinion on the nature of structures suspicious of being fungal elements. All too frequently surgical biopsies or autopsy tissues were not submitted for culture and a final diagnosis may rest on the proper interpretation of fungal elements in tissue sections. Therefore, a brief orientation would appear to be within the realm of "practical" mycology.

Once medical technologists and clinical microbiologists gain confidence in reviewing tissue sections, they will realize that fungi often do not appear substantially different in tissue sections than the forms seen in laboratory cultures incubated at 35°–37° C. Individuals who have had no formal training in the interpretation of stained tissue sections should not be confused by the background cellular reaction, but rather should concentrate on the fungal elements that are present. Yet, the basic types of inflammatory responses produced in human tissues in response to fungi are not that difficult to recognize with a little orientation. In fact, certain groups of fungi may be suspected based on the inflammatory reaction they produce, as presented in some detail by Ayers et al. (5).

With some variation, fungi produce one of the following four basic types of inflammatory tissue reactions:

**Purulent (suppurative)**
**Nonspecific chronic**
**Granulomatous**
**Necrotizing or inert**

Although virtually any fungus can produce one or more of these reactions depending upon the stage of inflammation and the immune status of the host, certain associations can be made.

**Purulent inflammation** is characterized by the accumulation of polymorphonuclear leukocytes (the same cells seen in the peripheral blood smear) at the site of infection, in confined areas known as **abscesses** (Plate 1.1*A*) or exude in the form of a fluid we visually observe as pus. This reaction is commonly seen in infections caused by pyogenic bacteria, such as the staphylococci; however, several fungi may also produce a purulent response. The yeast forms of *Blastomyces dermatitidis* (particularly in cutaneous lesions), which measure 8–

15 $\mu$m in diameter and typically show a single broad-based bud, are often seen within purulent abscesses (Plate 1.1*B*). The smaller (3–4 $\mu$m in diameter) endospores of *Coccidioides immitis* may also produce a purulent response at the site where they are released into the tissues (Plate 1.1*C*).

The broad, irregular, twisted, aseptate hyphae of *Zygomycetes* sp are also usually associated with a purulent response (Plate 1.1*D*). These hyphal forms may also be seen invading through vascular walls producing thrombosis (Plate 1.1*E*) with the potential of spread to distant organs. **Asteroid bodies** (Plate 1.1*F*), representing antibodies affixed to the surface of fungal elements, may be seen in mycotic infections, particularly in the subcutaneous infections caused by *Sporothrix schenckii.* The tissue response to infections with the Actinomycetes, filamentous bacteria belonging to the genera *Actinomyces* and *Nocardia,* is also usually purulent (Plate 1.1*G* and *H*).

If the acute phase reaction fails to remove the etiological agent, as is often the case with fungal organisms, the polymorphonuclear leukocytes may be replaced in time by dense accumulations of lymphocytes, plasma cells, macrophages, and varying numbers of eosinophils. This histological picture is seen in nonspecific chronic inflammation and may be seen during one of the inflammatory stages of virtually any fungus.

**Granulomatous inflammation** is a term used to describe discrete, nodular collections of large eosinophilic macrophages that often take on an epithelial appearance, and are thus referred to as epitheloid histiocytes (Plate 1.2*A*). They may appear singly, in clusters or aggregate to form multinucleated giant cells, often enveloping fungal elements within the common cytoplasm (Plate 1.2*B*). Epitheliod giant cells with nuclei distributed around the periphery are known as **Langhans' giant cells**. The aggregation of epitheloid macrophages, lymphocytes and plasma cells and Langhans' giant cells are the histological features of the classical tubercle, at one time considered diagnostic of tuberculosis. *Histoplasma capsulatum* and *C. immitis* are among the fungi that can be associated with this type of granulomatous inflammation.

*H. capsulatum* is a tiny (2–4 $\mu$m) yeast that is found intracellularly within epitheloid histiocytes (Plate 1.2*B*). The organisms may be difficult to see in routine hematoxylin and eosin (H & E) stained tissue sections but are readily visible in sections stained by the Gomori methenamine silver (GMS) technique (Plate 1.2*C*). The typical tissue form of *C. immitis* is a spherule, ranging in size from 10–60 $\mu$m or more in diameter. When young, spherules appear as empty shells (Plate 1.2*D*); when mature, they are filled with 2–4 $\mu$m in diameter endospores (Plates 1.1*C* and 3.2*G*). The granulomas produced by these organisms may progress to healing forming tiny fibrotic or calcified nodules.

Patients who are severely immunosuppressed often are incapable of eliciting a cellular response; **necrotizing or inert inflammation** is the result. The exudates are composed chiefly of edema, hemorrhage, deposition of fibrin and dead (necrotic) cells. *Cryptococcus neoformans* is one organism that often does not elicit a inflammatory response. This organism exists as a yeast form in tissue sections and can be recognized by yeast cells that vary in size between 3–10 $\mu$m and typically are surrounded by a thick capsule (Plate 1.2*E*).

Infections by *Aspergillus* sp and *Candida* sp are almost always opportunistic, occur in immunosuppressed hosts and often do not elicit a cellular response. *Aspergillus* sp appear as uniform, slender, dichotomously branching, septate hyphae (Plate 1.2*F*). In cases of cavitary fungus ball infections, fruiting heads may be seen in tissue sections (Plate 1.2*G*). *Candida* sp also produce **pseudohyphae** in tissues; however, in addition, produce budding yeast cells known as **blastoconidia** (Plate 1.2*H*). The pseudohyphae have regularly spaced points of constriction (resembling a link sausage), a feature by which they can be distinguished from the hyphae of *Aspergillus* sp which have parallel walls without constrictions.

The species of fungi briefly discussed in the preceding paragraphs are those most commonly found in specimens submitted to the surgical pathology laboratory and usually can be identified by the criteria outlined. Other fungal species are less commonly found in tissue sections; texts such as the color atlas by Chandler et al. (12) can be consulted to aid in making identifications. As mentioned above, fungal forms in tissues appear similar to those seen in laboratory cultures and often can be identified by individuals not formally trained in the interpretation of stained tissue sections.

### The Microbiology Laboratory

The major portion of this text is devoted to a discussion of the laboratory techniques and procedures required to recover and identify fungi from clinical specimens. Until recently, clinical microbiology laboratories have placed little emphasis on the recovery and identification of fungal etiological agents. As discussed by Thomson and Roberts (85), the levels of laboratory services devoted to mycology vary in different institutions. These authors make the plea that all clinical laboratories should at least provide cultural services for the recovery of fungi in clinical specimens and have the capability to recognize fungal agents when seen in microscopic mounts of clinical specimens.

There is a current renewed interest in clinical mycology resulting largely from the increased incidence of fungal diseases in patients with compromised host resistance and from the challenge resulting from working up samples sent out as part of various laboratory proficiency and accreditation programs. Physicians are more aware of the potential presence of mycotic disease in immunosuppressed patients and more frequently request the assistance of clinical laboratories. The following chapters include information designed to provide a basic and practical approach to the recovery and identification of the more commonly encountered fungal species. This will assist medical technologists and microbiologists to better serve the needs of the referring physicians.

# PLATE 1.1

## INFLAMMATORY TISSUE REACTIONS SEEN IN FUNGAL INFECTIONS

*A*. H & E-stained section of skin revealing purulent inflammatory reaction characterized by the formation of multiple intraepithelial microabscesses containing dense aggregates of polymorphonuclear leukocytes. Low power.

*B*. Histological tissue section through focus of purulent and subacute inflammatory exudate including large yeast cells, two of which show single buds (*arrows*), features consistent with yeast forms of *B. dermatitidis*. High power.

*C*. Histological section through focus of purulent inflammation at the site of a ruptured *C. immitis* spherule where endospores are being discharged into the tissue (*arrow*). High power.

*D*. H & E-stained tissue section through focus of purulent inflammation illustrating poorly staining fragments of ribbon-like, aseptate hyphae (*arrows*), characteristically seen in cases of invasive zygomycosis. High power.

*E*. Cross-section of a small arteriole illustrating fungal hyphae penetrating the wall and extending into the thrombosed lumen, as may be seen in cases of invasive Zygomycosis. High power.

*F*. Focus of purulent inflammation from an histological section of subcutaneous tissue illustrating an asteroid body surrounding a large budding yeast form, a picture seen in lesions caused by *S. schenckii*. High power.

*G*. Focus of purulent inflammation in histological section from a subcutaneous abscess illustrating dense aggregates of polymorphonuclear leukocytes surrounding an actinomycotic granule. Low power.

*H*. Gram-stained preparation of periphery of an actinomycotic granule illustrating delicate, deeply-staining, branching bacterial filaments. Oil immersion.

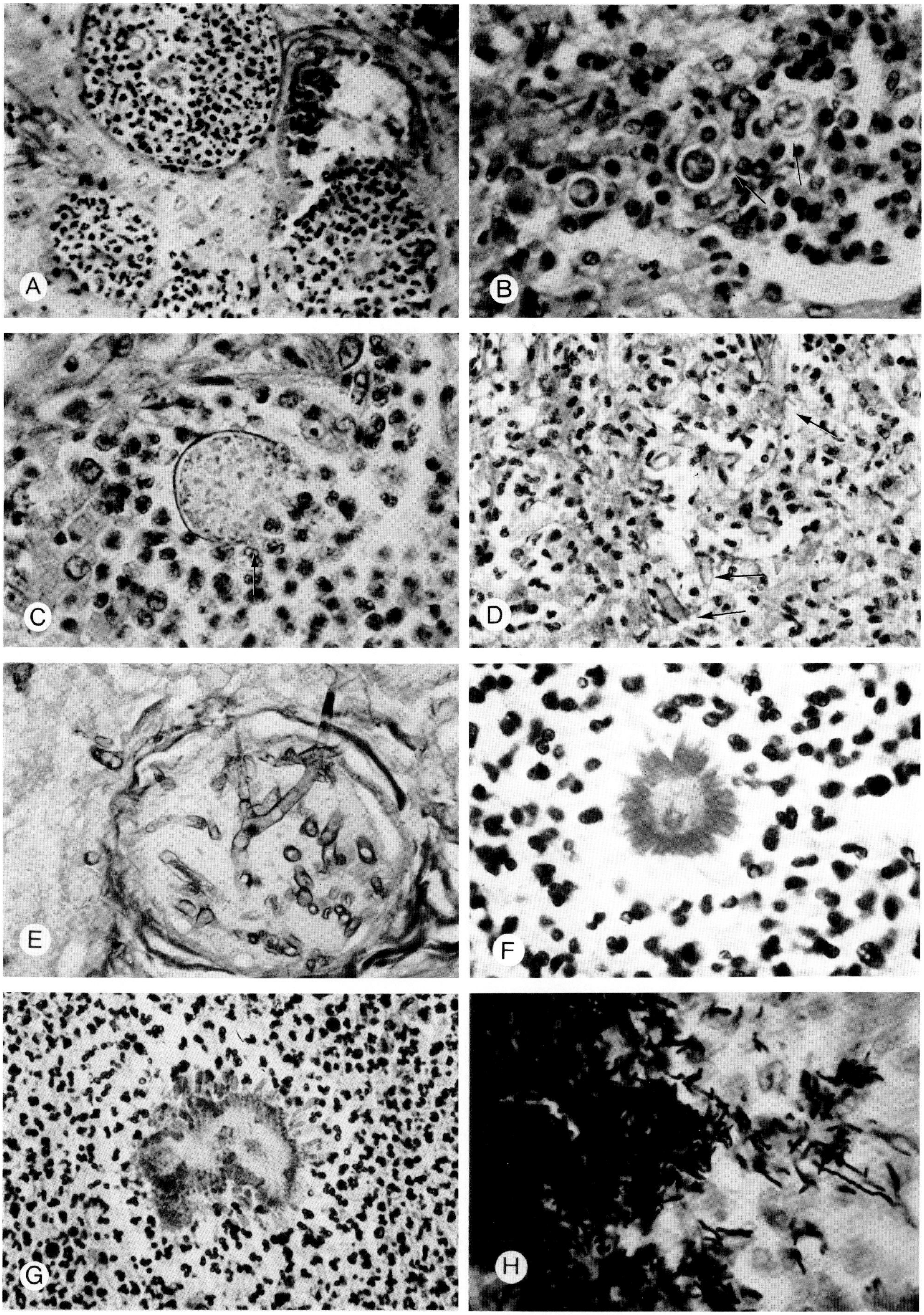
A
B
C
D
E
F
G
H

# PLATE 1.2

## INFLAMMATORY TISSUE REACTIONS SEEN IN FUNGAL INFECTIONS

*A*. Histological picture of granulomatous inflammation illustrating proliferation of histiocytic reticuloendothelial cells (*arrows*) and scattered small, round inflammatory cells. High power.
*B*. Demonstration of multinucleated giant cells, a form often seen in histological preparations of granulomatous inflammation. Note the presence of several intracytoplasmic pseudoencapsulated, 2–3 $\mu$m in diameter yeast cells characteristic of *H. capsulatum* (*arrows*). Oil immersion.
*C*. Methenamine silver-stained tissue section of granulomatous inflammation illustrating intracellular clusters of 2–3 $\mu$m yeast cells characteristic of *H. capsulatum.* Low power.
*D*. Focus of granulomatous inflammation demonstrating several immature spherules of *C. immitis*. Low power.
*E*. Focus of inert inflammation demonstrating aggregation of varying sized yeast cells widely separated by capsular material, a histological appearance characteristic of infection with *C. neoformans*. Note absence of inflammatory cells. High power.
*F*. Histological section through nidus of inert inflammation heavily infiltrated with dichotomously branching, regular, septate hyphae characteristic of invasive aspergillosis. High power.
*G*. Fruiting heads of *Aspergillus* sp as they are characteristically seen in histological sections of pulmonary fungus ball infections. High power.
*H*. Methenamine silver-stained section of a kidney from a patient with disseminated candidosis illustrating a nidus of intraglomerular pseudohyphae and blastoconidia characteristic of *Candida* sp. Note the absence of an inflammatory response. High power.

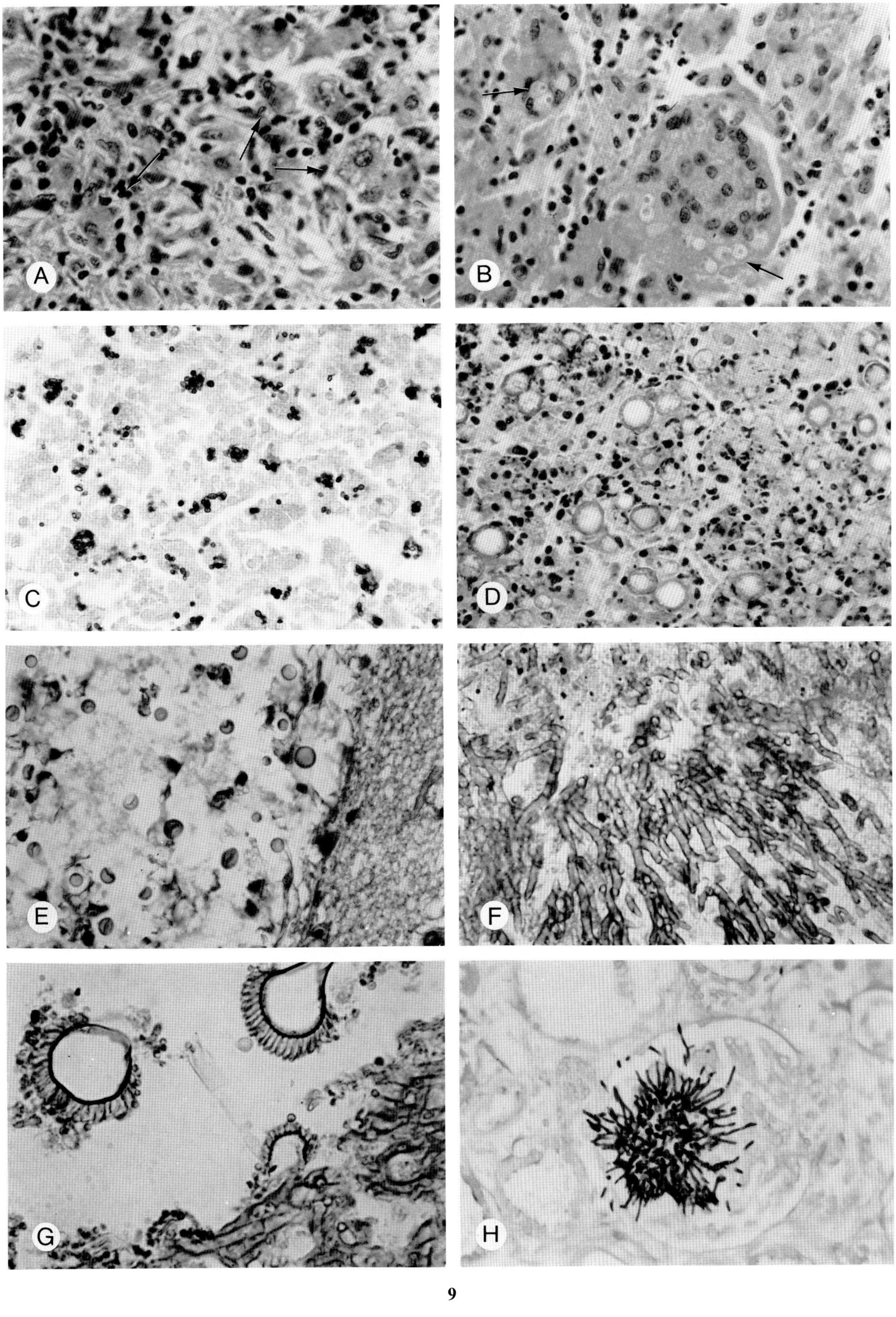
A
B
C
D
E
F
G
H

# Chapter 2

# SELECTION OF CLINICAL SPECIMENS FOR FUNGAL CULTURE

The diagnosis of mycotic infections is contingent upon the selection of the appropriate clinical specimen for culture. Table 2.1 presents the commonly encountered human mycotic infections and the specimens from which the etiological agents are most often recovered. It should be recognized that most mycotic infections begin in the lungs; therefore, respiratory specimens virtually always must be obtained. However, it should also be recognized that dissemination to other body sites often occurs and organisms can often be recovered from nonrespiratory secretions or tissues. *Histoplasma capsulatum*, for example, may also be recovered in disseminated cases from the bone marrow, blood, urine, and mucocutaneous lesions.

Thus, proper collection and rapid transport of clinical specimens to the clinical laboratory is of major importance for the recovery of the etiological agents of mycotic infections. Specimens may not only contain the causative agent but also a mixture of several bacteria or other fungi that may rapidly overgrow slower growing pathogenic fungi. Because overgrowth with these other microorganisms may occur, it is desirable to transport specimens with as little delay as possible.

Specimens should be transported in a sterile container that maintains or provides a moist environment; if necessary, sterile saline without a preservative can be added. Environmental conditions may also compromise the recovery rate; for example, *Blastomyces dermatitidis*, *Cryptococcus neoformans*, and *H. capsulatum* do not survive well in sputum that has been frozen or stored on dry ice for one or more days (46). If storage is necessary, specimens should be held at 4° C for no longer than 24 hours with the realization that some loss of viability may occur. Another recent study shows, however, that there may be some margin for error. Isolates of *H. capsulatum*, *B. dermatitidis*, *C. neoformans*, *Aspergillus* sp, *Sporothrix schenckii*, and *Nocardia asteroides* were still recovered from specimens that had been in transit to the laboratory for as long as 16 days, although rates of recovery were higher when subcultures were performed within 3 days (74). The rule of thumb here is, do not give up on the possibility of recovering a fungal isolate just because a specimen has been delayed in transit; however, strive to process specimens as soon after collection as possible.

## CRITERIA FOR REJECTION OF UNSUITABLE SPECIMENS

A protocol should be established in each laboratory that lists the criteria by which unacceptable specimens can be rejected. This list should be published and posted in hospital or clinic nursing stations and circulated to out-of-house physicians who refer specimens. A suggested list of criteria for specimen rejection is shown in Table 2.2.

The specimen must be adequate. Spit is not sputum; a sparsely inoculated dry swab is next to worthless and a few casually collected skin scales or randomly plucked hairs may not be representative of the disease.

The specimen container must be properly labeled and accompanied by a request slip that provides the name, room number or address, and hospital or social security number of the patient, the name of the referring physician, the source of the culture and information on the clinical condition and the organism suspected. Special media or more rigid culture procedures may be required to recover the species of certain pathogenic fungi, such as *H. capsulatum, Coccidioides immitis* and *B. dermatitidis.*

**Table 2.1.** Most Likely Recovery Sites for Common Pathogenic Fungi

| | Suspected Pathogens | | | | | | | | | | | |
|---|---|---|---|---|---|---|---|---|---|---|---|---|
| Specimens | Dermatophytes | *B. dermatitidis* | *C. immitis* | *H. capsulatum* | *Paracoccidioides brasiliensis* | *S. schenckii* | *Aspergillus* sp | *Candida* sp[a] | *C. neoformans*[b] | *N. asteroides*[c] | *Pseudallescheria boydii* | Zygomycetes[d] |
| Lower respiratory tract | | X | X | X | X | X | X | X | X | X | X | X |
| Blood | | | X | X | | X | | X | X | X | | |
| Bone | | X | X | | | | X | X | X | | X | X |
| Hair, nails | X | | | | | | | | | | | |
| Bone marrow | | | | X | X | X | | | X | X | | |
| Brain | | X | X | X | | | X | X | X | X | X | X |
| CSF | | | X | X | | X | | X | X | | | |
| Ear | | | | | | | X | | | | | |
| Eye | | | | X | | X | X | X | X | | X | X |
| Liver/spleen | | | X | X | X | | | | X | X | | |
| Nose/nasal sinus | | | | | | X | X | | | | X | X |
| Prostate | | X | | | | | | | | | | |
| Skin/mucous membrane | X | X | X | X | X | X | X | X | X | X | X | X |
| Subcutaneous tissue | | | X | | | X | X | | | X | X | X |
| Synovial | | | X | | | X | | X | | X | | |
| Urine | | X | X | X | | | | X | X | X | | |
| Multiple systemic sites | | X | X | X | | X | X | X | X | X | X | X |
| Vagina | | | | | | | | X | | | | |

[a] *Candida* sp are most frequently recovered from respiratory secretions and vaginal mucous membranes; however, their clinical significance from these sites is questionable.

[b] More than one site shall be cultured in suspected cases. Serological testing for capsular antigen is recommended on CSF and in serum in disseminated cases.

[c] An acid-fast bacterium commonly recovered on fungal media.

[d] The clinically significant Zygomycetes include the genera *Mucor* and *Rhizopus*.

**Table 2.2.** Criteria for Rejection of Mycology Culture Specimens

| Criteria for Rejection | Action | Criteria for Rejection | Action |
|---|---|---|---|
| No identification on container or discrepancy between patient information on request form and container label | Return to sender for resolution | Dried out swab or insufficient material | Notify physician to request a new specimen if warranted by clinical condition; if a new specimen is indicated and cannot be obtained by noninvasive culture techniques, procurement of biopsy specimens may be necessary |
| Sputum for bacterial culture with 25 epithelial cells per low power field as observed in a Gram's-stained smear | If a lower respiratory infection is suspected, the presence of large numbers of squamous cells in the sputum sample indicates contamination with oral secretions; this is not necessarily a criterion for rejection in that pathogenic fungi can be recovered in the face of contamination, particularly if selective fungal culture media is used; each case must be judged on its own merits | 24-hour urine or sputum collections received for mycobacteria ("TB") and/or fungi | Notify physician that 24-hour collections are unacceptable and request submission of 3 single-voided early morning urine samples or 3 consecutive early morning, freshly expectorated sputa |
| | | Adequate material submitted in improper container; that is, evidence of leakage, lack of sterility, bed pan, etc. | Call physician or nursing service and request resubmission of sample in appropriate container as detailed in the laboratory protocol manual |

## SPECIFIC SPECIMEN SOURCES

### Lower Respiratory Tract

**Sputum** should be collected from a deep cough early in the morning soon after the patient rises. The mouth should be vigorously rinsed with water (not a proprietary gargle) immediately before collecting the specimen to minimize the concentration of oropharyngeal contaminants. Twenty-four hour collections of sputum are unacceptable, not only because it is inconvenient for the patient to collect the sample, but also because samples usually become overgrown with bacteria or saprophytic fungi.

When the patient cannot expectorate sputum, saline or **Isuprel** may be introduced into the bronchial tree through the use of intermittent positive breathing equipment, which generally produces deep spasms of coughing. Whether spontaneously produced or induced by nebulization, 10 to 15 ml of the patient's sputum should be collected in a sterile, screw-capped container, tightly sealed, properly labeled, and delivered promptly to the laboratory (see Table 2.2).

Transtracheal aspiration may be helpful in obtaining specimens from patients who are debilitated and cannot produce sputum or who are too ill to withstand the induction procedure. Bronchoscopy biopsies may be helpful in the diagnosis of invasive pulmonary mycoses, particularly if the infiltrates or other lesions are located more peripherally in one or more lobes of the lungs. Samples should be collected in sterile containers without preservatives that can be easily sealed and forwarded immediately to the laboratory for processing.

### Genitourinary Tract

**Urine** samples may be collected via the midstream, clean-catch technique or with a sterile needle and syringe from the soft-rubber connector of an indwelling urinary catheter. Do not obtain urine samples from a collection bag or bed pan. The recovery of fungal elements is optimum from the first morning sample, particularly following a 12-hour water fast. Twenty-four hour urine samples are unacceptable for the same reasons as those mentioned for the collection of 24-hour sputum samples.

Urine samples should be processed as rapidly as possible after collection. Specimens kept at room temperature for longer than one hour may not be useful for the detection of some dimorphic fungi associated with disseminated mycotic infections; hold specimens at 4° C if a delay in processing is anticipated.

Obtaining cultures from the **uterine cervix** or **vaginal canal** is of questionable value because of the common presence at these sites of commensal yeasts that may overgrow the culture medium.

### Cutaneous Specimens

Cutaneous samples can be obtained by scraping skin scales with the edge of a surgical blade or microscope slide, by plucking hairs with forceps or by or clipping infected nails and placing the samples into a sterile Petri dish. Collection of these specimens into a clean envelope for transport via mail is also acceptable.

The areas to be sampled should be wiped first with a cotton swab saturated with 70% alcohol to remove surface bacterial contaminants. Skin lesions should be sampled from the erythematous, peripheral, actively growing margins of typical "ringworm" infections. An open Petri dish can be conveniently pressed into the skin immediately below the area to be sampled and skin scales can be flaked off using the side of a surgical blade or the edge of a sterile microscope slide and collected in the bottom of the plate. In sampling infected nails, scrape away the superficial portions with the side of a surgical blade before collecting a deeper sample from which recovery of fungal elements will be more likely.

Potentially infected hairs can be plucked with a pair of surgical forceps. The use of a Wood's ultraviolet lamp may be helpful in illuminating those hairs that should be selected if the infection is due to one of the dermatophyte species that produces fluorescence (*Microsporum audouinii*, for example).

### Subcutaneous Specimens

Suppurative lesions of the deep skin and subcutaneous tissue, where pus may be loculated within abscesses or is exuding from deep sinus tracts, aspiration with a sterile needle and syringe should be attempted. If immediate inoculation of the specimen into an appropriate culture medium is not possible, the material should be placed into an anaerobic transport tube. If a swab is used for collection of material, it should be extended into the depths of the wound without touching the adjacent skin margins. Both anaerobic and aerobic cultures should be performed, the former being necessary to recover the anaerobic branching filamentous bacteria belonging to the genus *Actinomyces*.

### Cerebrospinal Fluid

**Cerebrospinal fluid (CSF)** is usually collected by the physician who performs the routine lumbar puncture technique. Most commonly, three separate sterile tubes of CSF are collected: the first tube is used for determining the concentration of various chemical constituents, the second for performing cell counts, and the third for culture. The rationale for reserving the third tube for culture is that theoretically any contaminating bacteria introduced into the needle while transmigrating the skin will have washed into the first two tubes. Approximately 1–5 ml of fluid are commonly available for culture. Once received in the laboratory, CSF specimens should be processed promptly; if this is not possible, the samples should not be refrigerated, rather they should be left either at room temperature or placed into a 30° C incubator since the protein and carbohydrate-rich fluid itself is an adequate medium to maintain the viability of the organism.

### Blood

The laboratory should be informed by the physician if fungal septicemia is suspected because special media are necessary for the optimum recovery of fungi. Numerous blood culture systems are available; however, all systems must be vented to atmospheric air and incubated at 30° C to maximize the rate and time of recovery of fungal organisms. In low volume laboratories where the few requests for fungal blood cultures may not warrant maintaining a stock of special blood culture bottles, it is permissible to transfer approximately 0.5–1.0 ml of buffy coat (prepared by centrifuging 5–10 ml of blood) to the surface of brain-heart infusion agar (supplemented with 5% sheep blood). This inoculum can then be spread over the surface of the agar with a sterile glass rod or inoculating loop. The use of Sabouraud's dextrose agar should be avoided since many of the dimorphic fungi require growth supplements that are not included in Sabouraud's agar.

The recovery of fungal organisms from blood cultures can be improved if a **biphasic bottle** containing 60 ml of brain-heart infusion broth and a slant of brain-heart infusion agar is used (see Appendix II). Ten milliliters of blood should be added to this bottle, a sterile cotton, plugged needle placed in the rubber stopper to serve as a vent to atmospheric air, and the bottle incubated in an upright position at 30° C. Bottles are examined daily and the blood-broth mixture is flooded over the agar surface to provide a subculture. Cultures are incubated for 30 days. If only a broth-containing bottle is used, it may be difficult to detect early growth of fungi since many species do not produce turbidity.

The **Isolator** (E.I. du Pont Inc., Wilmington, DE) is a lysis centrifugation system that has been found to improve the recovery of fungi and other microorganisms from blood cultures (9). The Isolator utilizes a tube that contains components that lyse leukocytes and erythrocytes and also inactivate complement and perhaps other antibacterial agents in plasma (Fig. 2.1). Once lysed, the cells release the microorganisms contained within them, and the centrifugation step in the procedure serves to concentrate the organisms in the blood sample. This concentrate is inoculated onto the surface of appropriate culture media. Major advantages of this system are that fungi are detected at a higher rate and the time to positivity is decreased markedly when compared to the performance of the biphasic brain-heart infusion bottle. **We recommend that the**

**Isolator system be used when processing blood cultures from patients with suspected fungal septicemia.**

### Tissue Specimens

Tissue biopsies are obtained either in the operating room or in the physician's office or clinic. Samples should be placed into a 4 × 4 sterile surgical gauze moistened with sterile saline. If the specimen cannot be delivered immediately to the laboratory, the gauze should be placed into a container with a screw-cap lid to prevent drying during transport. Samples submitted in formalin are usually not suitable for culture, although it may still be possible to recover fungal organisms from the centers of larger specimens if the time of fixation has been short. Care should also be taken not to place biopsy samples into **"saline for injection"** solutions which contain antimicrobic substances.

Tissue specimens should be homogenized before inoculating them onto appropriate culture medium. This may be accomplished with the use of a Teflon grinder or a mortar and pestle using sterile sand or other suitable nontoxic abrasive to effect more complete homogenization. Currently available from Tekmar, Inc. in Cincinnati, Ohio is an automated tissue processing device called the **Stomacher** (Fig. 2.2) which more gently massages the sample, an important consideration for the recovery of some of fragile hyphal forms that may not withstand the more rigorous treatment in a tissue grinder. Larger samples should be first cut into 1- to 2-mm cubes with a sterile surgical blade before grinding. Optimally, 0.5–1.0 ml of tissue homogenate

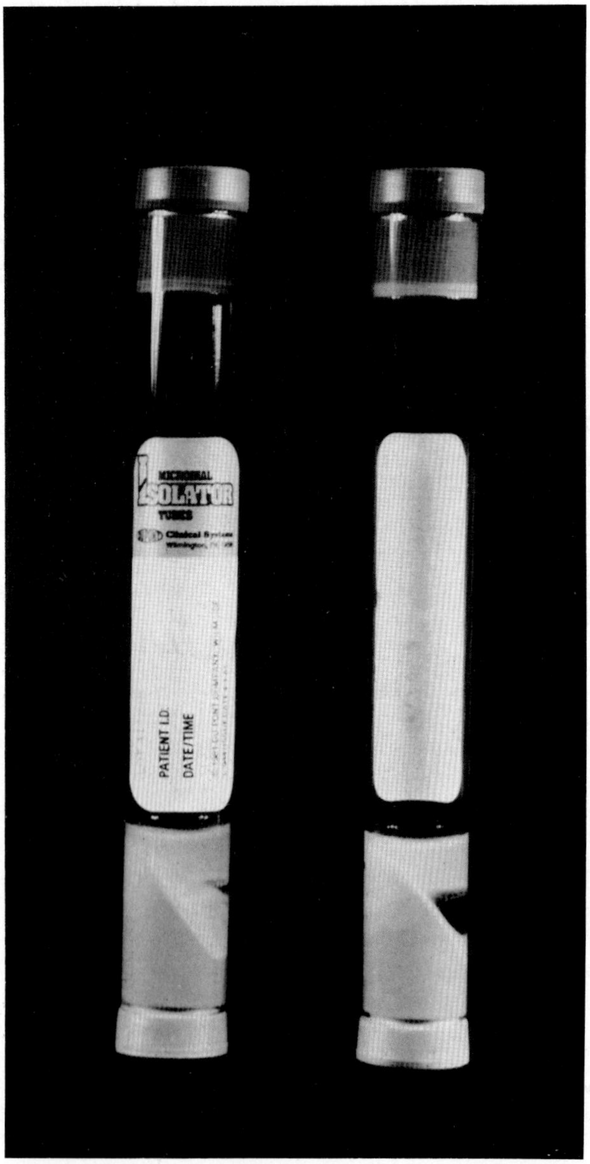


**Figure 2.1.** Dupont Isolator tubes containing a lysing agent and other components that remove cells from the blood sample and inactivate plasma complement and certain antibiotics.

should be spread over the surface of appropriate culture media and incubated in atmospheric air at 30° C.

### Bone Marrow

Aspiration or core needle biopsy samples are obtained by the physician using standard sterile techniques. With the aspiration technique, the initial material is generally used for making bone marrow smears; subsequently, an additional 3–5 ml of marrow and blood are removed and placed into a sterile vial containing 0.5 ml of 1:1000 heparin. Core biopsies obtained for culture can be placed directly onto the surface of appropriate media. Cultures are incubated at 30° C.

### Body Fluids and Exudates

Fluids that accumulate as part of an infectious or inflammatory process are called **transudates** if the specific gravity is less than 1.013 or, **exudates** if the specific gravity is greater. Samples are usually obtained by aspiration with a sterile needle and syringe. If the volume of fluid is large, such as the collection of thoracic (paracentesis) or abdominal (ascitic) fluids, the use of

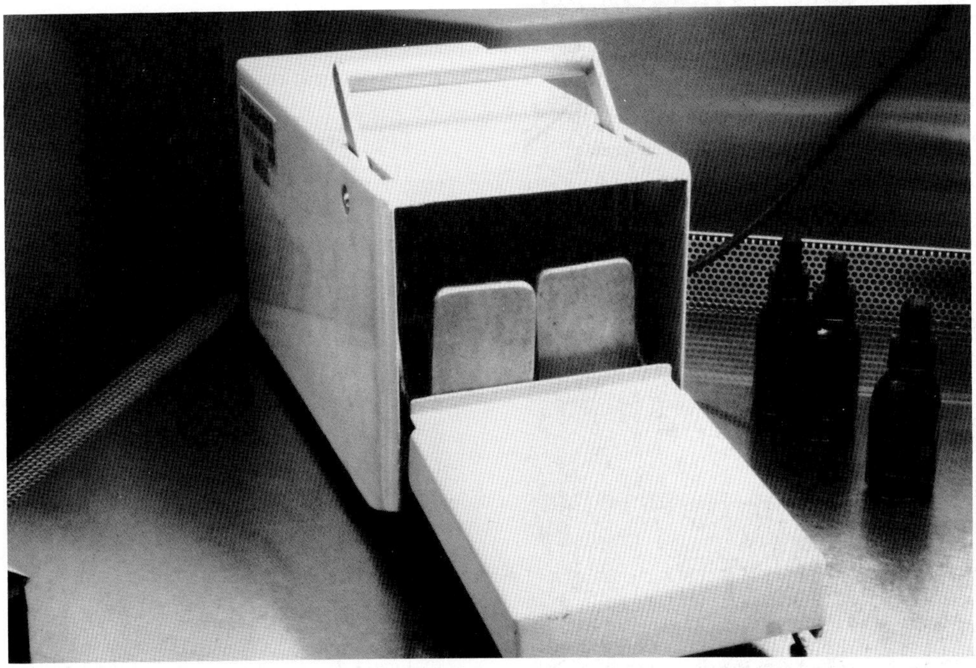

**Figure 2.2.** Tekmar Stomacher looking at the paddles through the open door. A plastic bag containing diluent and the specimen are placed in front of the paddles and the door is closed. During operation of the device, the back and forth action of the paddles gently massages the specimen and extracts any microorganisms that may be present.

sterile gallon-size containers may be necessary. Since transudates and exudates often contain sufficiently high concentrations of thromboplastin-like substances and fibrin precursors, about 10 ml of 1:1000 sterile heparin may be added to the container prevent clotting of the sample. The fluid should be allowed to settle at room temperature for one hour or more. The supernatant can then be poured off and 10–50 ml aliquots of the bottom settled fluid removed for centrifugation. About 0.5–1.0 ml of centrifuged sediment can then be placed on the surface of appropriate media.

## Summary

Table 2.3 summarizes the methods for collecting specimens for fungal culture as described in the above paragraphs. If fungal organisms are recovered from one or more sites in the absence of overt clinical signs or symptoms, the patient should be followed closely with periodic complete physical examinations and appropriate X-ray or serological studies since latent disease may be present in a subclinical or early progressive stage of development.

**Table 2.3.** Methods for Collecting Specimens for Fungal Culture[a]

| Specimen | Methods |
|---|---|
| Lower respiratory tract | Expectorated sputum: obtain three successive early morning samples<br>Induced sputum: services of a respiratory therapist are usually elicited<br>Gastric aspirate: aspirate about 10 ml of gastric fluid from a nasogastric tube; early morning specimens are optimal<br>Transtracheal aspiration:, Bronchoscopy:, Lung biopsy:, Pleural biopsy: Procedures usually performed by a physician; see that appropriate specimen tube and transport container are supplied |
| Blood | Put 10 ml of venous blood into biphasic blood culture bottle or use Dupont Isolator Tube. |
| Bone marrow | Aspirate 5–10 ml of bone marrow-blood mixture and place in sterile vial containing about 0.5 ml of 1:1000 dilution of sterile sodium heparin |
| CSF | Obtain as much fluid for culture as possible; usually the third tube in the collection sequence is submitted for culture |
| Eye | Obtain drainage material with a swab; this procedure usually optimal for recovery of bacteria; corneal scraping, intraocular fluid aspiration, or biopsy may be required for recovery of fungi |
| Hair | Select fluorescent hairs using a Wood's lamp or those in an infected area and remove with sterile forceps; removal of hair shaft is necessary |
| Nails | Decontaminate the surface of the nail with 70% alcohol, scrape away the outer portion and obtain scrapings from the deeper infected areas |
| Nose/nasal sinuses | Exudate may be cultured; however, often is negative for fungi; use necrotic material or obtain biopsy for optimal recovery of fungi |
| Skin/mucous membranes | In suspected dermatophyte infection, decontaminate skin with 70% alcohol before obtaining superficial skin scales with side of surgical blade or edge of a microscope slide; make direct smears of purulent exudates of subcutaneous or mucous membrane lesions. If fluid is loculated, aspirate with sterile needle and syringe |
| Tissue biopsy | Using aseptic technique, obtain two biopsies from the lesion if possible; one to include the wall or adjacent normal tissue, the other from the center of the lesion |
| Urine | Obtain three consecutive, early morning samples collected by the routine midstream technique |

[a] It may be necessary to culture multiple specimens from a single source to isolate the etiological agent.

# Chapter 3

# DIRECT MICROSCOPIC EXAMINATION OF CLINICAL SPECIMENS

We encourage the direct microscopic examination of specimens submitted to the laboratory for fungal culture, particularly those obtained from immunosuppressed hosts because an immediate presumptive diagnosis can often be made. Same-day diagnosis may serve as a guide to chemotherapy based on identifying fungal elements in direct preparations of fluid sediments, in exudative material, or in stained frozen sections of tissues. Smears can be prepared and special stains for fungi applied. The periodic acid-Schiff (PAS) stain is valuable but more time-consuming than the preparation of unstained mounts as described below. Fungal elements may also be recognized in Papanicolaou's (PAP)-stained cytological preparations and with a newly described stain, calcofluor white (29a).

**Direct mounts** are easy to prepare and fungal elements can often be detected. These mounts are made by mixing a small portion of the specimen to be cultured in 2 or 3 drops of water, physiologic saline, or 10% potassium hydroxide on a microscope slide. Potassium hydroxide is used for specimens such as skin, nail scrapings and infected hairs, or for other specimens such as sputum to clear out background debris that may be confused with fungal elements. The use of KOH is not necessary for specimens such as cerebrospinal fluid where background debris is minimal. KOH mounts are prepared as shown in the opposite column.

The contrast between unstained fungal elements that may be present and the background mounting fluid can be accentuated by narrowing the iris diaphragm to reduce the amount of incident light; or, use of phase contrast microscopy can greatly enhance visualization of organisms. Alternately, **calcofluor white** in a final concentration of 0.05% (see Appendix II) can be added to 10% KOH and the specimen examined under ultraviolet or blue light fluorescence. Fungal elements will take on a bright **apple green fluorescence** (29a).

**KOH Mount**

1. Place a drop of 10% KOH containing 10% glycerin on a microscope slide and mix a small quantity of the material to be examined (skin or nail scrapings, hairs, etc).
2. Gently pass the slide through a low flame of a Bunsen burner to facilitate clearing (do not boil).
3. Place a cover slip (18 × 18 mm) over the drop and let the mount stand at room temperature for approximately 30 minutes if the specimen is thick or viscous.
4. Examine microscopically for the presence of hyphae or other fungal structures (see Plate 3.5*A*).

Frozen sections of tissue biopsies are prepared using a cryostat fitted with a microtome knife. Thin, 4–6 μm sections are cut, alcohol-fixed on a microscope slide and stained with hematoxylin and eosin (H&E) and/or Gomori methenamine silver (GMS). Impression smears prepared from bloody biopsies or bone marrow can be stained with Wright-Giemsa; smears of purulent exudates in which one of the bacterial species belonging to the genus *Actinomyces* may be present should be stained with the Gram's method and an acid-fast technique if *Nocardia* sp is suspected. If an interpretation cannot be made from the frozen section material, next-day diagnosis may be possible by examining permanent tissue sections stained with H & E, GMS, PAS, Gridley's fungus (GF) stain, or specialized stains such as Mayer's mucicarmine (capsular mucopolysaccharide of *Cryptococcus neoformans*).

*C. neoformans* may also be recognized by using the **India ink technique** which aids in the detection of the characteristic capsules when present. **Nigrosin** (available from Harleco, a division of Hartmann-Ledded Co., Philadelphia, PA) can be substituted for India ink. The mount is prepared as follows:

**India Ink Preparation**

1. Place a drop of India ink or nigrosin on a microscope slide.
2. Transfer a small portion (loopful) of spinal fluid sediment and mix thoroughly in the drop of India ink or nigrosin.
3. Place a cover slip (18 × 18 mm) over the drop and examine under the low and high power objectives of a microscope.
4. The presence of yeast cells, some exhibiting "pinching" or budding with a narrow-neck attachment, surrounded by a thick capsule, usually easy to see because it is not penetrated by the ink particles, is diagnostic of *C. neoformans* (see Plate 3.3*A*, *B*, and *C*).

*Cryptococcus* yeast forms may be mistaken for inflammatory cells such as leukocytes and erythrocytes or artifacts such as talc crystals (contamination from the powder applied to sterile gloves). Usually these forms are easy to distinguish with experience; however, on occasion, it may be necessary to prepare Wright- or Giemsa-stained smears to recognize the yeast cells of *C. neoformans* definitively.

The India ink or nigrosin preparation is not recommended for the diagnosis of cryptococcal meningitis since less than 50% of proven cases exhibit positive results. The diagnosis is best accomplished by use of the cryptococcal latex test for antigen.

When cells resembling *C. neoformans* are seen in specimens other than cerebrospinal fluid, India ink or nigrosin can be used to demonstrate capsules. This is particularly helpful when small budding yeast cells are observed in sputum specimens and capsules are not evident by bright field or phase contrast microscopy.

The selection of stains depends upon the experience and personal preferences of surgical pathologists and upon the type of fungus suspected. The GMS, PAS, and GF stains are based on the principle that, in the presence of chromic or periodic acid, adjacent hydroxyl groups of the polysaccharides in fungal cell walls are oxidized to aldehydes. Aldehydes, in turn, form a black methenamine silver nitrate complex in the GMS stain or a red-purple reaction with the Schiff reagent in the PAS or GF stains (12).

As mentioned in Chapter 1, it is important for microbiologists to remember that the fungal forms seen in direct mounts of clinical materials will appear identical to those seen in stained tissue sections, except for distortion that may occur from the inflammatory response to the host. These same forms will also be seen in laboratory cultures that have been incubated at 35°C; therefore, it should be possible for those not formally trained in the examination of stained smears or tissue sections still to be able to identify fungal elements that may be present.

The following pages will include several series of black and white photomicrographs portraying the appearance of fungal elements in direct mounts, smears, and in tissue sections stained with a variety of stains, as they appear with brightfield, phase contrast, and fluorescence microscopy. Each series of photographs will portray one of the potentially pathogenic fungal species that are most commonly encountered in clinical practice. The following outline will be used in sequencing the fungal species to be presented:

1. Species virtually always producing yeast forms in tissues:

   *Blastomyces dermatitidis* (Plate 3.1*A–D*).
   *Paracoccidioides brasiliensis* (Plate 3.1*E–H*).
   *Histoplasma capsulatum* (Plate 3.2*A–D*).
   *Cryptococcus neoformans* (Plate 3.3*A–E*).
   *Sporothrix schenckii* (Plate 3.3*F–H*).

2. Species producing spherules in tissue:

   *Coccidioides immitis* (Plate 3.2*E–H*).

3. Species virtually always producing hyphal or filamentous forms in tissue:

   *Aspergillus* (Plate 3.4*A–D*).
   Zygomycetes (Plate 3.4*E–H*).
   Dermatophytic fungi (Plate 3.5*A–E*).
   Agents of subcutaneous mycoses (Plate 3.5*F–H*).
   Actinomycetes including *Nocardia* sp (filamentous bacteria) (Plate 3.6*E–H*).

4. Fungal species producing both yeast and hyphal or pseudohyphal forms in tissue:

   *Candida* species (Plate 3.6*A–D*).

# PLATE 3.1

## DIRECT EXAMINATION OF CLINICAL SPECIMENS

***Blastomyces dermatitidis***

*A*. Phase contrast photomicrograph of a purulent respiratory tract specimen containing a heavy concentration of inflammatory cells and a budding yeast cell of *B. dermatitidis.* Note the broad based nature of the bud, the double contoured wall, the intracytoplasmic material and large size (12 μm). High power.
*B*. Photomicrograph of PAP-stained cytology smear of respiratory secretions illustrating a typical budding cell of *B. dermatitidis* within a cluster of benign epithelial cells. High power.
*C*. Typical yeast cells of *B. dermatitidis* in methenamine silver-stained tissue section. Note that some of the yeast cells exhibit single buds with a broad base. High power.
*D*. Photomicrograph of auramine-rhodamine-stained tissue exudate exhibiting typical yeast cell of *B. dermatitidis.* High power.

**Summary of Diagnostic Features**

Large budding yeasts
8–15 μm in size (range 2–30 μm)
Appear to have a "double contoured" wall
Buds connected by a broad base
Several buds may remain attached
Intracytoplasmic contents evident
Tissue reaction purulent (early) or granulomatous (late)
Must differentiate from *Paracoccidioides brasiliensis*

***Paracoccidioides brasiliensis***

*E*. Yeast forms of *P. brasiliensis* as seen by phase contrast in a saline wet mount of respiratory secretions. Note the multiple buds surrounding the parent cell. These yeast cells are similar in size to those of *B. dermatitidis*, ranging from 8–15 μm in diameter. High power.
*F*. Methenamine silver-stained yeast cell of *P. brasiliensis* illustrating the multiple buds, giving the appearance of a "mariners wheel." High power.
*G*. Methenamine silver-stained tissue section illustrating multiple 12–15 μm in diameter yeast cells of *P. brasiliensis* exhibiting typical mariner's wheel appearance. High power.
*H*. Methenamine silver-stained tissue section illustrating many smaller and irregular-sized yeast forms of *P. brasiliensis*. The smaller forms have some resemblance to the yeast cells of *H. capsulatum*; the larger and more irregular-sized forms may be confused with the encapsulated yeast cells of *C. neoformans.* The identification of multiple buds is the chief identifying characteristic, as illustrated by the dark-staining cell in the right upper corner. High power.

**Summary of Diagnostic Features**

Large, thin-necked, budding yeasts
8–40 μm in size (range 5–60 μm)
Multiple buds surround parent cell resembling a "mariner's wheel"
Smaller buds formed singly may resemble *H. capsulatum*
Tissue reaction purulent (early) or granulomatous (late)
Must differentiate from *B. dermatitidis.*

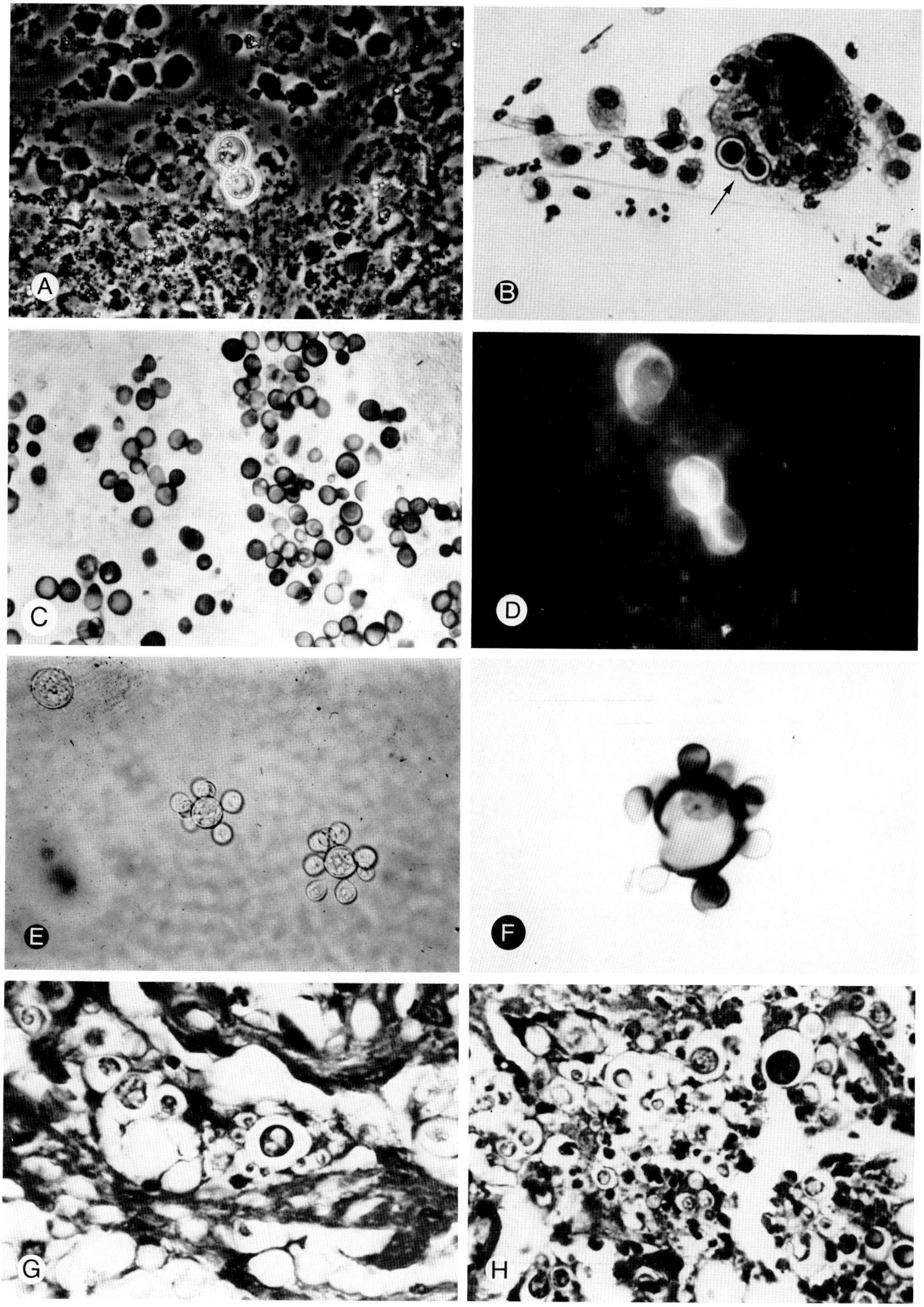
A
B
C
D
E
F
G
H

# PLATE 3.2

## DIRECT EXAMINATION OF CLINICAL SPECIMENS

***Histoplasma capsulatum***

*A.* H&E-stained tissue section illustrating large histiocytic reticuloendothelial cells containing numerous 2–4 μm pseudoencapsulated intracellular yeast forms of *H. capsulatum.* Oil immersion.

*B.* Wright-stained preparation of bone marrow smear containing numerous 2–3 μm pseudoencapsulated yeast cells of *H. capsulatum*, some appearing to be liberated from a ruptured reticulum cell (*arrow*). The appearance of the yeast cells here closely resembles the aflagellar leishmanial forms of *Leishmania donovani.* Oil immersion.

*C.* Photomicrograph of calcofluor white-stained tissue exudate illustrating numerous intracellular fluorescing yeast cells of *H. capsulatum.* Oil immersion.

*D.* Methenamine silver-stained tissue section showing intracytoplasmic clusters of yeast cells of *H. capsulatum.* High power.

**Summary of Diagnostic Features**

Small budding yeasts
Spherical to oval
Intracellular; may have pseudocapsule
2–5 μm in size
Single buds with narrow bases
Difficult to detect in unstained clinical specimens
Tissue reaction granulomatous with giant cells and caseation necrosis
Must differentiate from *Candida glabrata*, nonbudding endospores of *C. immitis* and smaller forms of nonencapsulated *C. neoformans*

***Coccidioides immitis***

*E.* Phase contrast photomicrograph of tissue exudate showing immature spherules of *C. immitis* devoid of endopores. High power.

*F.* Purulent tissue exudate illustrating two spherules of *C. immitis* lying adjacent so as to resemble a budding yeast cell of *B. dermatitidis.* High power, bright field photomicrograph.

*G.* Methenamine silver-stained tissue section containing *C. immitis* spherules in varying stages of development. Note that the larger, more mature spherules contain endospores. To the left, note rupture of some of the spherules with release of endospores into the tissue. Low power.

*H.* Methenamine silver-stained section of tissue showing *C. immitis* endospores and rudimentary hyphae as they may appear in a pulmonary cavity.

**Summary of Diagnostic Features**

Large spherules
10–60 μm in size (range 20–200 μm)
May contain nonbudding endospores 2–5 μm in size
Immature spherules vary greatly in size and are often devoid of endospores
Septate hyphae may be found in cavitary lesions
Spherules lying side-by-side may resemble *B. dermatitidis*; endospores may resemble *H. capsulatum* but do not exhibit budding
Tissue reaction granulomatous but purulent early

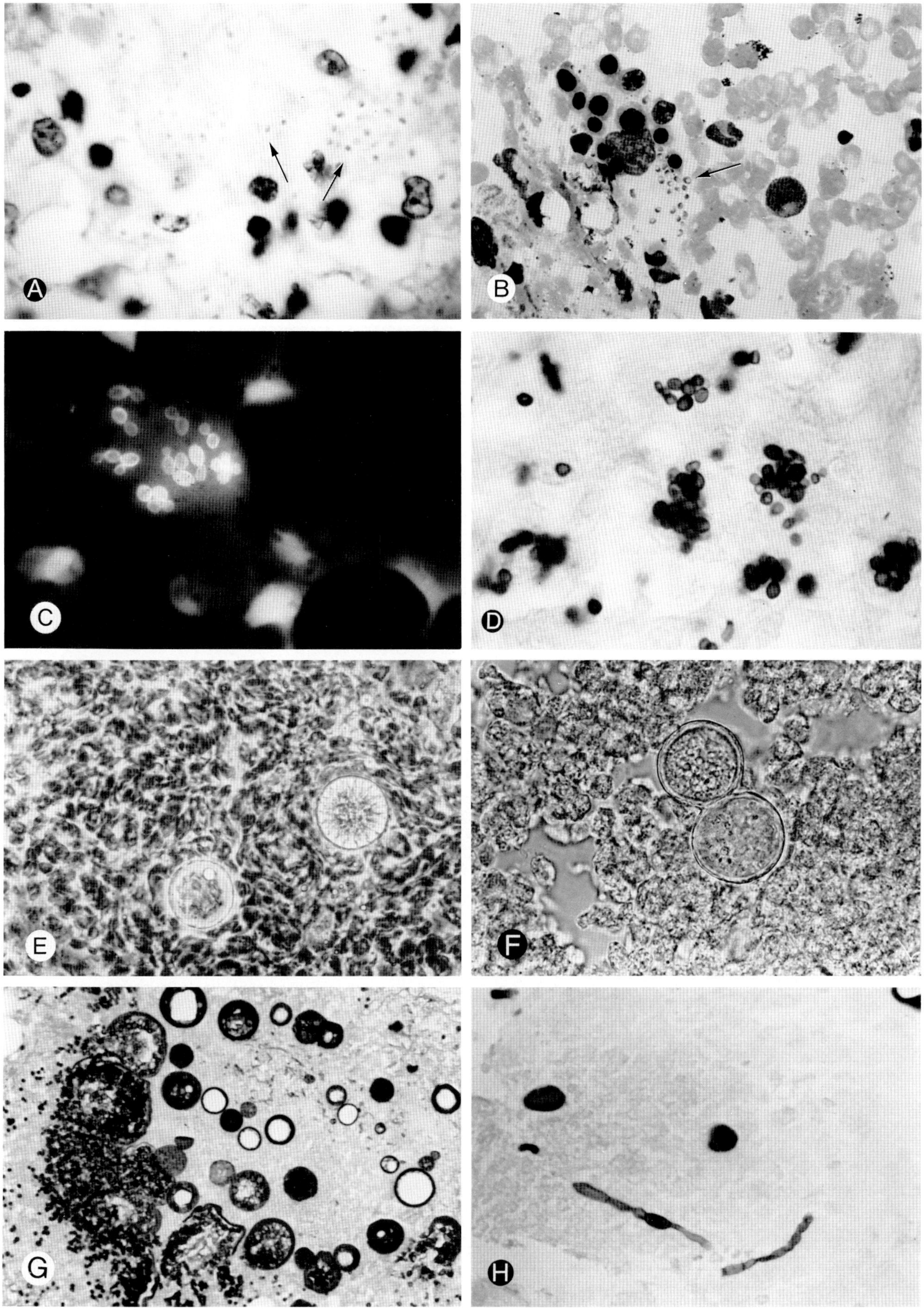
A
B
C
D
E
F
G
H

# PLATE 3.3

## DIRECT EXAMINATION OF CLINICAL SPECIMENS

***Cryptococcus neoformans***

*A*. Phase contrast photomicrograph of tissue exudate of yeast cells of *C. neoformans*, each surrounded by a narrow capsule (*arrows*). High power.

*B*. India ink preparation showing numerous cells of *C. neoformans.* Note the spherical nature of the cells which vary in size and are surrounded by large capsules. Note the budding form in the lower right corner of the field and the daughter cell connected by a narrow, pinched attachment. High power.

*C*. Phase contrast photomicrograph of respiratory secretions illustrating a typical yeast cell of *C. neoformans* surrounded by a large capsule and exhibiting a single pinched-off bud. High power.

*D*. Methenamine silver-stained tissue section illustrating yeast cells of *C. neoformans* appearing widely separated by capsular material. Note the wide size variation of the yeast cells. High power.

*E*. PAP-stained preparation of respiratory secretions illustrating a pseudohyphal form of *C. neoformans* surrounded by capsular material. Although *C. neoformans* rarely produce pseudohyphae in cultures, they may be seen in tissue sections or in secretions. Oil immersion.

**Summary of Diagnostic Features**

Budding yeast cells
2–15 $\mu$m in size
Yeast cells spherical, vary in size
Usually encapsulated
"Pinched off" buds; sometimes multiple
Encapsulated pseudohyphae sometimes present
Tissue response granulomatous, necrotizing or inert
Must differentiate from *H. capsulatum*; cells lying singly side-by-side may resemble *B. dermatitidis*

***Sporothrix schenckii***

*F*. Gram-stained smear of tissue exudate illustrating numerous yeast cells of *Sporothrix schenckii.* Note elongated forms, the so-called "cigar bodies" (*arrow*). High power.

*G*. Methenamine silver-stained yeast forms of *S. schenckii* demonstrating both oval and elongated forms. Oil immersion.

*H*. H&E-stained tissue section of purulent abscess illustrating an asteroid body within a sea of polymorphonuclear leukocytes. Asteroid bodies are composed of antibody protein surrounding a microorganism, an occurrence not uncommon in lesions caused by *S. schenckii.* Note the budding yeast form of *S. schenckii* within the center of the asteroid body. Oil immersion.

**Summary of Diagnostic Features**

Small budding yeasts
2–6 $\mu$m in size
Oval to elongated (resembling cigar-forms)
May exhibit multiple budding
Difficult to detect in clinical specimens
Tissue reaction purulent or granulomatous, with asteroid body formation
Must differentiate from *H. capsulatum*

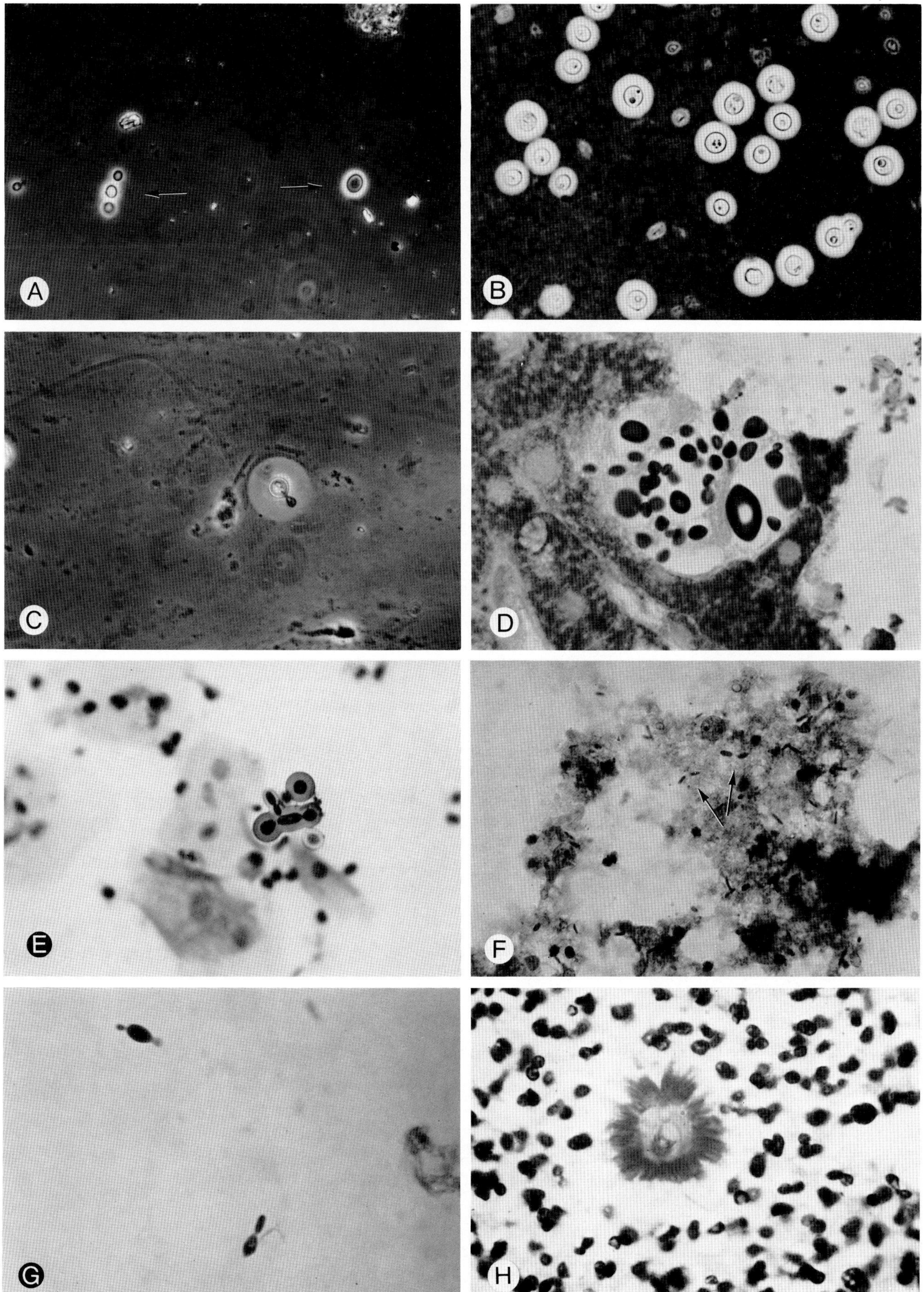
A
B
C
D
E
F
G
H

# PLATE 3.4

## DIRECT EXAMINATION OF CLINICAL SPECIMENS

**_Aspergillus_ species**

*A*. Septate hyphae of *Aspergillus* in respiratory secretions as viewed by phase contrast. Note the regular size of the hyphae and the dichotomous branching. High power.
*B*. PAP-stained smear of respiratory secretions showing branching, septate hyphae of *Aspergillus* sp. High power.
*C*. Photomicrograph of methenamine silver-stained tissue section illustrating septate, dichotomously branching hyphae of *Aspergillus* sp. High power.
*D*. H&E-stained tissue section of material from pulmonary cavity (fungus ball) containing short, septate hyphae and fruiting heads of *Aspergillus* sp (*arrows*). High power.

**Summary of Diagnostic Features**

Septate hyphae, lie parallel
Dichotomous branching at 45° angles
3–6 μm in size, but some larger up to 12 μm
Larger hyphae may resemble those of the Zygomycetes
Tissue reaction granulomatous, necrotizing or inert in immunosuppressed hosts
Must differentiate from *Candida* sp, Zygomycetes and other saprobic fungi; i.e. *Pseudallescheria boydii.*

**Zygomycetes**

*E*. Phase contrast view of tissue exudate containing broad, irregularly branching aseptate hyphae of a Zygomycete. High power.
*F*. PAP-stained preparation of respiratory secretions illustrating non-septate hyphae of Zygomycetes. Note the variation in width of the hyphae, smaller of ones of which may be in the size range of *Aspergillus* sp. High Power.
*G*. H&E-stained tissue section revealing short, irregular ribbon-like hyphae of Zygomycetes. High power.
*H*. Methenamine silver-stained tissue section illustrating several mature sporangia containing sporangiospores. These forms may resemble endospore-containing spherules of *C. immitis.* The fruiting sporangia of Zygomycetes as illustrated here are most commonly found in fungus ball lesions within cavitary lesions. High power.

**Summary of Diagnostic Features**

Large ribbon-like hyphae, often twisted and fragmented
10–20 μm in size; range 3–25 μm
Branching irregularly
Usually nonseptate (rare septations may be observed)
Smaller hyphae may resemble those of *Aspergillus* sp; but usually do not have parallel walls
Tissue reaction suppurative with abscess formation
Must differentiate from *Aspergillus* sp and other saprobic fungi

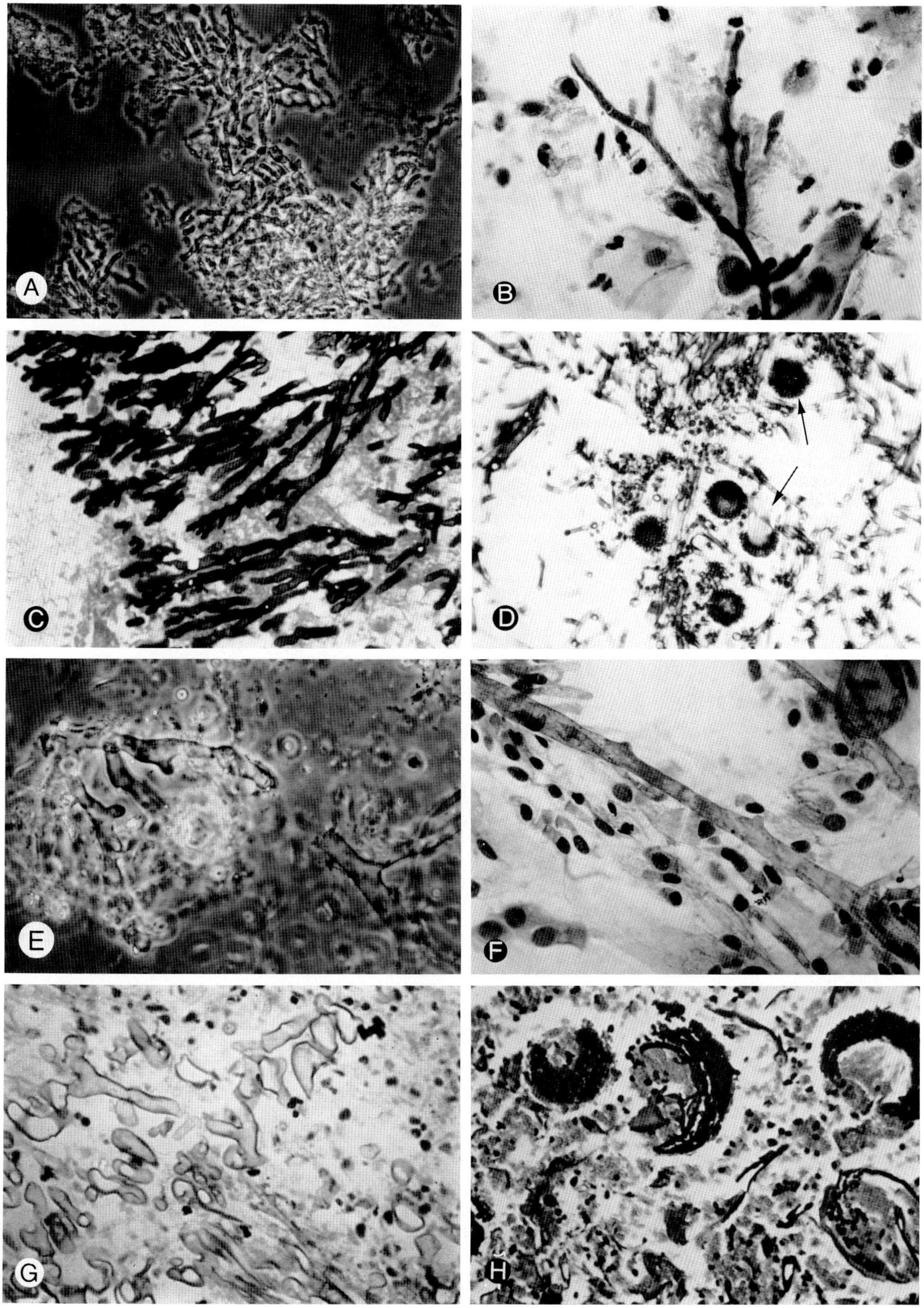
A
B
C
D
E
F
G
H

# PLATE 3.5

## DIRECT EXAMINATION OF CLINICAL SPECIMENS

***Dermatophytic Fungi***

*A.* KOH preparation of skin scales observed under phase contrast illustrating typical hyphal segment of a dermatophytic fungus with characteristic arthroconidia formation. High power.
*B.* Split frame of hairs infected with a dermatophytic fungus. The left frame reveals endothrix arrangement of hyphal segments splitting into arthroconidia; the right frame reveals ectothrix aggregation of conidia in a mosaic pattern. High power.
*C.* Phase contrast view of skin scales containing hyphae and budding yeast cells of *Malassezia furfur* from a case of tinea versicolor. Note that the yeast cells retain their integrity and are not flattened, even though lying adjacent to one another in compact clusters. High power.
*D.* PAS-stained preparation of skin scales revealing delicate hyphae and budding yeast cells characteristic of *M. furfur.* High power.
*E.* Histological section of superficial cutaneous biopsy revealing invasion of stratum corneum with short hyphal segments characteristic of infection with a dermatophyte (*arrow*). High power.

**Summary of Diagnostic Features**

Small septate hyphae
1–2 μm in size
No branching evident
Arthroconidia formation common
Hair invasion internal (endothrix) or external (ectothrix)
Tissue reaction—hyperkeratosis and acanthosis. Dermal reaction usually mild and mononuclear in type
Must differentiate from *M. furfur* that has short hyphal elements (usually can be differentiated by the clusters of round, thick-walled cells resembling "spaghetti and meatballs")

**Agents of Subcutaneous Mycoses**

*F.* H&E-stained section of subcutaneous abscess illustrating a eumycotic granule from a patient with mycetoma. Low power.
*G.* Methenamine silver-stained section of subcutaneous mycetoma granule revealing eumycotic nature with hyphal segments and characteristic swollen cells at periphery. High power.
*H.* H&E-stained section of subcutaneous granuloma from a patient with chromoblastomycosis showing varying sized sclerotic bodies characteristic of infection with *Fonsecaea pedrosoi.* The spherical bodies resemble "copper pennies." High power.

**Summary of Diagnostic Features**

Sulfur granules present—eumycotic mycetoma
1–5 mm in size
Various colors
Granules contain septate hyphae, 2–6 μm in size
Swollen cells at periphery of granule
Granules surrounded by eosinophilic material
Dematiaceous hyphae seen with *Exophiala jeanselmei*
Tissue reaction in suppurative and granulomatous
Splendore-Hoeppli phenomenon evident

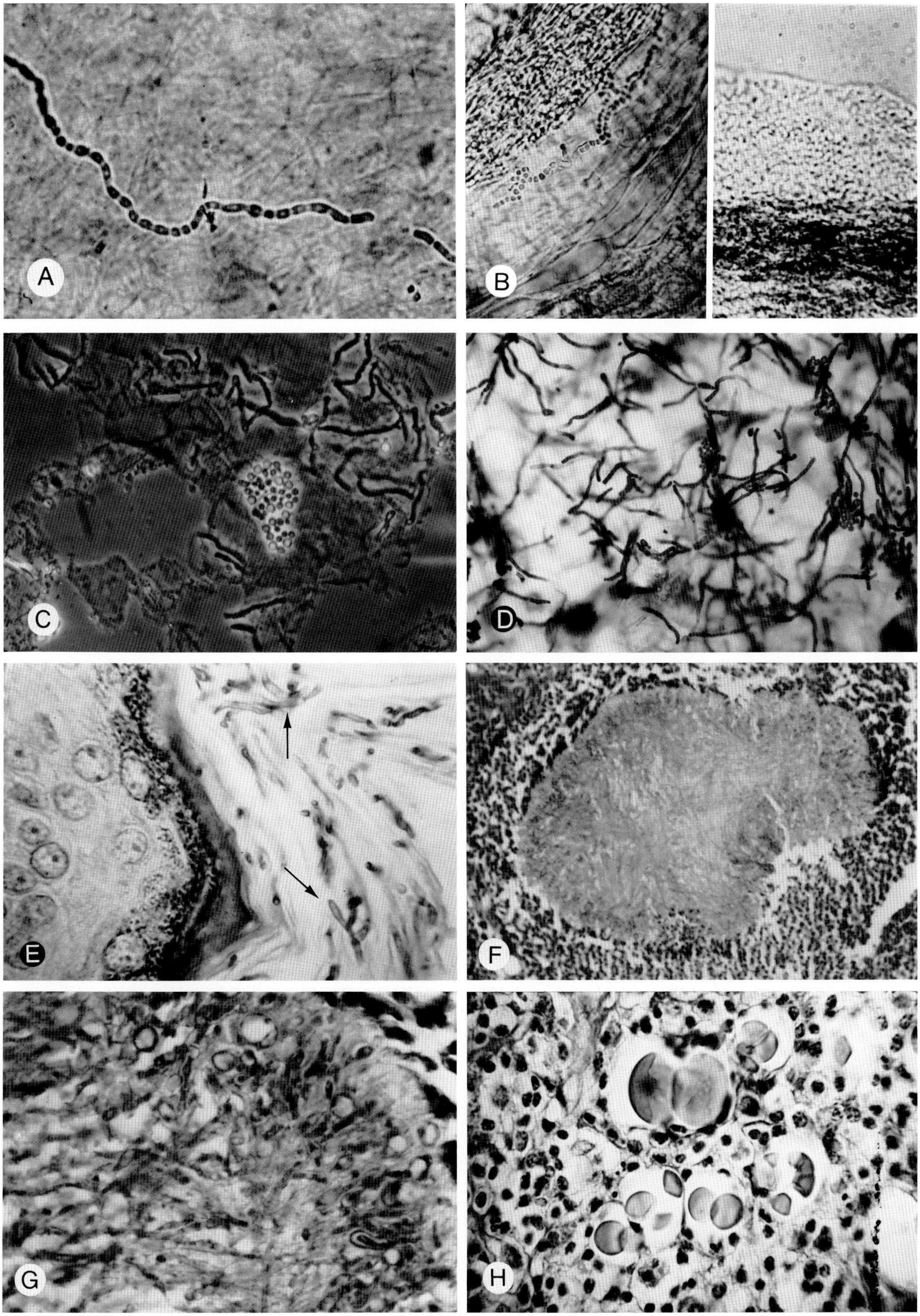
A
B
C
D
E
F
G
H

# PLATE 3.6

## DIRECT EXAMINATION OF CLINICAL SPECIMENS

***Candida* species**

*A.* Gram-stained preparation of urinary sediment revealing pseudohyphae and blastoconidia characteristic of *Candida* sp. High power.
*B.* Urine sediment observed under phase contrast illustrating pseudohyphae and blastoconidia characteristic of *Candida* sp. High power.
*C.* Methenamine silver-stained tissue section from a case of disseminated candidosis illustrating a nidus of blastoconidia (best seen centrally) and pseudohyphae radiating from the periphery. Note the absence of an inflammatory response indicating immune suppressed status of the host. Low power.
*D.* PAS-stained section of a granuloma from a young leukemic child with disseminated candidosis illustrating short segments of pseudohyphae, blastoconidia, and several large, spherical, deeply staining chlamydospores (*arrows*) (an unusual occurrence). Oil immersion.

**Summary of Diagnostic Features**

Budding yeast cells, oval or round
3–4 μm in size
Pseudohyphae also present, showing regular points of constriction, resembling link sausages
Hyphae, when present are septate
Tissue reaction purulent (early) and granulomatous later. Necrotic or inert in immunosuppressed hosts
Must be differentiated from *Aspergillus* sp, *Trichosporon* sp (with blastoconidia) and *Geotrichum* sp (with arthroconidia)

**Actinomycetes Including *Nocardia* species**

*E.* Potassium hydroxide preparation of subcutaneous exudate from a patient with an actinomycotic mycetoma illustrating the microscopic appearance of a "sulfur" granule. Note clubbing of filaments at the periphery of the granule. High power.
*F.* H&E-stained preparation of subcutaneous abscess revealing a full field of neutrophils surrounding a cluster of actinomycotic granules. High power.
*G.* Gram-stained preparation of actinomycotic granule revealing peripherally placed dense aggregates of deeply staining delicate filaments. Oil immersion.
*H.* Acid fast-stained preparation of a subcutaneous exudate revealing delicate, branching, beaded filaments characteristic of *Nocardia asteroides.* High power.

**Summary of Diagnostic Features**

Sulfur granules present—actinomycotic mycetoma
1–5 μm in size
Various colors; usually yellow or white
Granules contain delicate, branching filaments, 0.1–1.0 μm in size
Granules surrounded by eosinophilic protienaceous material
Tissue reaction—suppurative, Splendore-Hoeppli phenomenon evident

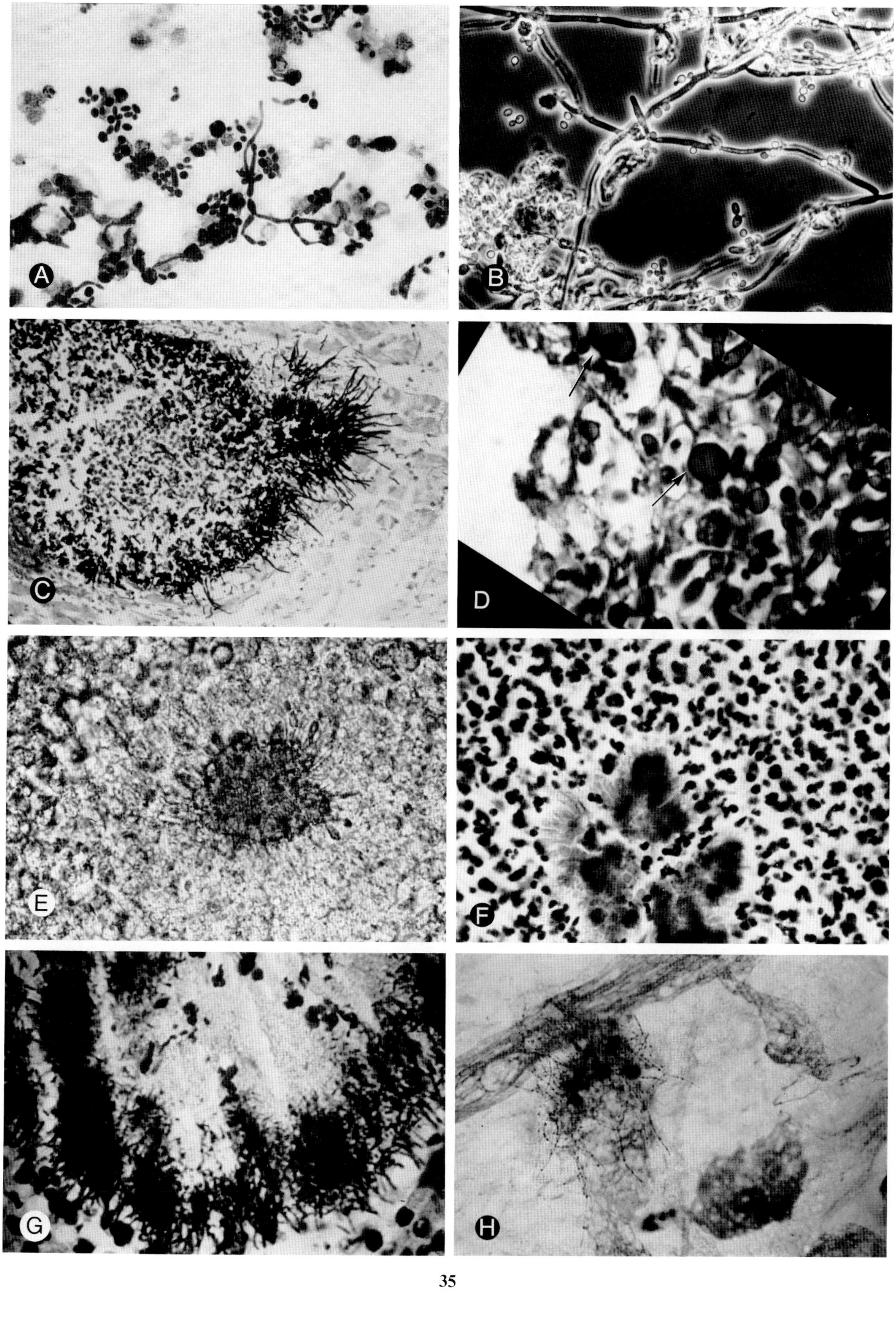
A
B
C
D
E
F
G
H

# Chapter 4

# PROCESSING AND CULTURING OF CLINICAL SPECIMENS

After proper collection, specimens should be placed into an appropriately labeled sterile container and transported to the laboratory. Processing should not be delayed to ensure the recovery of fastidious fungi that may be present and to prevent the overgrowth of bacteria or rapidly growing saprobic fungi. Although the culturing of specimens falls within the realm of the clinical laboratory, physicians should be aware of the common procedures performed in the laboratory to which cultures are referred so that additional information can be provided to ensure maximum recovery of fungi from unusual cases.

Table 4.1 provides a guide for the selection of common media useful for the recovery of fungi from specific anatomic sites and should be used in conjuncton with the recommendations given in the paragraphs that follow. Although there are similarities in the processing techniques for specimens from various sources, there are differences which are also outlined in the following discussion.

## PROCESSING OF SPECIMENS

### Respiratory Secretions

The recovery of pathogenic fungi from sputum may be increased by selecting the purulent parts of the sample for culture, adding a mucolytic agent (an equal quantity of 0.5% N-acetyl-L-cysteine), and thoroughly stirring the mixture prior to centrifugation to obtain a concentrated sediment. It is important that the digestion mixture used for the treatment of sputum for the recovery of mycobacteria is not inadvertently used since the NaOH contained in the mycobacterial digestion mixture inhibits most of the fungal forms that may be present (72). It must be realized that contaminating bacteria are also concentrated by this technique and may overgrow nonselective fungal culture media. Liquefaction is not required for bronchial specimens but may be helpful for the processing of tracheal aspirates depending upon the viscosity of the secretions. No data are available to show that liquefaction and concentration of clinical specimens enhance the recovery rate of fungi.

Approximately 0.5 ml of sputum or sediment, or an equal quantity of tracheal or bronchial secretions should be inoculated onto the appropriate media listed in Table 4.1. The inoculum should be spread over the surface of the agar either with a sterile glass rod, with a sterile, wide-bore pipette, or with an inoculating loop. An initial streak is first made down the center of the agar from one margin of the plate to the other; then, the plate is rotated 90° and streaked at right angles until the surface is covered. It is important to note that media containing antibacterial drugs should be selected in addition to antibiotic free media since common bacterial flora may interfere with the recovery of some strains of pathogenic fungi (see Table 4.1).

### Cerebrospinal Fluid

Three to five milliliters of cerebrospinal fluid (CSF) are optimal for the recovery of fungi; however, lesser volumes are often received and should be processed. Fungal elements can be concentrated either by centrifugation or by the use of a 0.45 $\mu$ membrane filter. In the centrifugation technique, the entire specimen is transferred to a sterile conical tube and spun at low speed (approximately 1500 × G) for 15 minutes. Contrary to popular belief, *Cryptococcus neoformans* is not adversely affected by centrifugation at this speed and organisms will remain viable. The supernatant should be kept for serological examination for fungal antigens or antibodies when required. One drop of sediment can be used for making an India ink preparation. The remaining sediment should be resuspended in approximately 2 ml of CSF, mixed, and 4 or 5 drops placed onto separate areas on the surface of an appropriate

**Table 4.1.** Fungal Culture Media: Indications for Use

| Primary Recovery Media | |
|---|---|
| **Media** | **Indications for Use** |
| Brain-heart infusion agar | Primary recovery of saprobic and pathogenic fungi |
| Brain-heart infusion agar with antibiotics | Primary recovery of pathogenic fungi exclusive of dermatophytes |
| Brain-heart infusion biphasic blood culture bottles | Recovery of fungi from blood |
| Dermatophyte test medium | Primary recovery of dermatophytes, recommended as screening medium only |
| Inhibitory mold agar | Primary recovery of pathogenic fungi exclusive of dermatophytes |
| Mycosel or mycobiotic agar | Primary recovery of dermatophytes |
| Sabouraud's 2% dextrose agar | Primary recovery of saprobic and pathogenic fungi (not recommended) |
| SABHI agar | Primary recovery of saprobic and pathogenic fungi |
| Yeast-extract phosphate agar | Primary recovery of pathogenic fungi exclusive of dermatophytes |
| **Differential Test Media** | |
| **Media** | **Indications for Use** |
| Ascospore agar | Detection of ascospores in ascosporogenous yeasts such as *Saccharomyces* sp. |
| Casein agar | Identification of *Nocardia* sp and *Streptomyces* sp. |
| Cornmeal agar with Tween 80 and trypan blue | Identification of *Candida albicans* by chlamydospore production. Identification of *Candida* by microscopic morphology |
| Cottonseed conversion agar | Conversion of dimorphic fungus *Blastomyces dermatiditis* from mold to yeast form |
| Czapek's agar | Recovery and differential identification of *Aspergillus* sp. |
| Niger seed agar | Identification of *C. neoformans* |
| Nitrate reduction medium | Detection of nitrate reduction in confirmation of *Cryptococcus* sp. |
| Potato dextrose agar | Demonstration of pigment production by *Trichophyton rubrum*; preparation of microslide cultures |
| Rice medium | Identification of *Microsporum audouinii* |
| Trichophyton agars 1–7 | Identification of members of *Trichophyton* genus |
| Tyrosine agar | Identification of *Nocardia* sp. and *Streptomyces* sp. |
| Urea agar | Detection of *Cryptococcus* sp.; differentiate *Trichophyton mentagrophytes* from *Trichophyton rubrum*; detection of *Trichosporon* sp. |
| Xanthine Agar | Identification of *Nocardia* sp. and *Streptomyces* sp. |
| Yeast fermentation broth | Identification of yeasts by determining fermentations |
| Yeast nitrogen base agar | Identification of yeasts by determining carbohydrate assimilations |

culture medium. Since CSF is unlikely to be contaminated with organisms other than the etiological agent, it is unnecessary to use an inhibitory medium for culture.

Concentration of the CSF specimen by filtration is recommended, particularly for volumes that exceed 2 ml. Using a syringe with a **Swinnex adapter** fitted with a 0.45 $\mu$m membrane filter, spinal fluid is filtered under

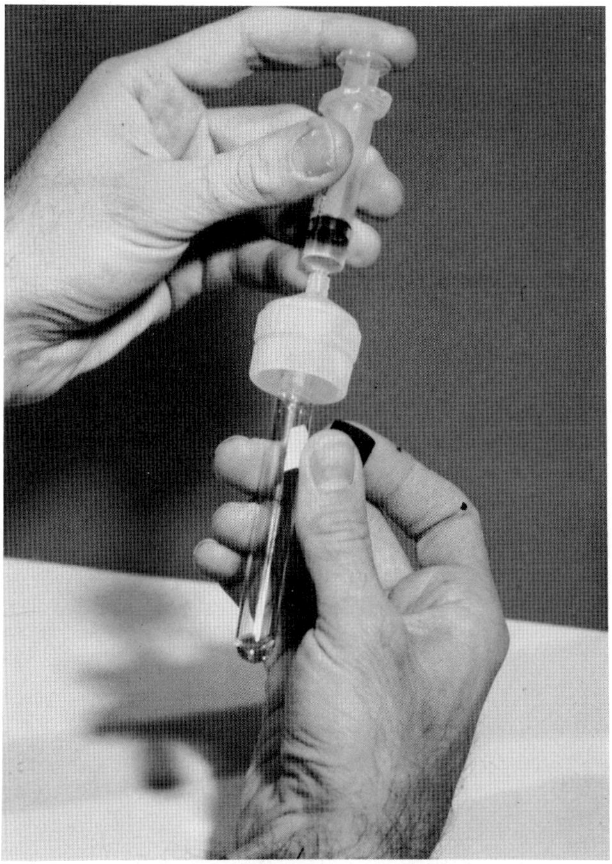

**Figure 4.1.** Swinnex filter device by which the fluid to be cultured is expressed from a syringe through a 0.45-$\mu$m micropore membrane on which any microorganisms that may be present are trapped.

pressure into a sterile tube (Fig. 4.1). The filtrate should be retained to perform serological tests if necessary. Following filtration, the membrane filter is removed and the side containing the sediment is placed in contact with the surface of appropriate agar medium from the list in Table 4.1. Each day the membrane filter should be lifted and placed at alternate sites on the agar surface and the exposed inoculation sites examined for the presence of growth (Fig. 4.2).

### Urine

Urine specimens should be processed as rapidly as possible after collection. Specimens should be concentrated by centrifugation prior to media inoculation. Approximately 0.5 ml of sediment should be inoculated to the surface of both inhibitory and antibiotic-free media and streaked following the procedure outlined

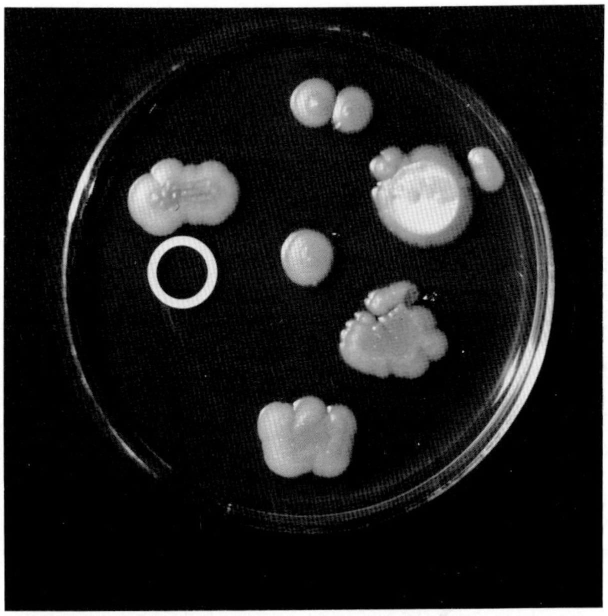

**Figure 4.2.** Surface of agar upon which the millipore filter had been placed and moved to multiple locations after successive periods of incubation. Note growth of colonies where the filter pad had been placed.

above for sputum culturing. The quantitation of fungal colonies recovered from urine cultures is still controversial and is of questionable benefit.

### Tissues and Other Body Fluids

Tissue specimens should be processed before inoculation onto appropriate culture media. They may first be minced or ground with a mortar and pestle or tissue grinder, using a small amount of water or saline. A small amount of sterile sand or other nontoxic abrasive agent added to the liquid aids in homogenizing the tissue. As much homogenate as possible should be transferred to appropriate culture media. If it is suspected that tissue fragments may contain a Zygomycete, the specimen should be minced rather than ground since the aseptate (coenocytic) hyphae present in the tissue may be destroyed, thereby compromising the recovery of viable forms in culture. As briefly mentioned in Chapter 2, an automated tissue processing device called the **Stomacher** (Tekmar, Inc, Cincinnati, OH) can be used since it tends to squeeze rather than tear the tissue (see Fig. 2.1). Alternatively, small bits of minced tissue can be inoculated with sterile forceps directly to the agar by submerging small fragments beneath the surface (growth of some fungi is enhanced under slightly anaerobic conditions).

Body fluids with a volume of 200 ml or more should be allowed to settle for 2 or 3 hours. The resulting sediment can then be placed in one or more 50-ml centrifuge tubes after the supernatant is decanted. Centrifuged sediments of each tube should be combined and at least 0.5 ml should be inoculated to the surface of appropriate culture media.

**Figure 4.3.** Mayo Clinic biphasic fungal-blood culture bottle containing brain-heart infusion broth bathing a brain-heart infusion agar slant.

### Blood Culture of Fungi

In small volume laboratories where the low frequency of requests for fungal blood cultures does not permit having special media on hand, it is acceptable to inoculate approximately 0.5 ml of heparinized blood directly onto the surface of an appropriate agar medium. These can be incubated aerobically at 30° C.

For the improved recovery of fungi from blood specimens, a **biphasic bottle** containing 60 ml of brain-heart infusion broth and a slant of brain-heart infusion agar should be used (Fig. 4.3). Ten ml of blood are inoculated into the bottle, that is vented with a sterile cotton-plugged needle and incubated vertically for 30 days at 30° C. Bottles are examined daily for visible evidence of growth. After each examination, the biphasic bottles are gently mixed so that the blood-broth mixture is flooded over the agar surface to allow any organisms present to implant on the surface of the agar slant. Bottles are held for 30 days before discarding as negative. Bone marrow aspirates can be treated similarly to blood samples as described above.

As discussed previously, we recommend the use of the Du Pont **Isolator** for the optimal recovery of fungi from blood cultures (Fig. 2.1). In the Mayo Clinic evaluation of the Isolator previously mentioned (9), not only is the detection rate of fungemia significantly increased but the time for cultures to become positive is also substantially shortened. Using the Isolator, most cultures of yeasts and yeast-like organisms have become positive within 3 days while those of *Histoplasma capsulatum* are often positive as soon as 8 days (these times are double or more when using conventional broth systems). A comparison of the detection times of fungi recovered by the Isolator and biphasic BHI bottles is presented in Table 4.2. Note that the mean detection time for the Isolator was 2.12 days compared to 4.90 days for BHI bottles. Thus, the Isolator can be recommended as the method of choice for recovering fungi from blood cultures.

### Skin Scrapings, Hair, and Nails

As many skin scales, nail fragments, or hairs as are available should be placed on agar culture media and portions submerged beneath the surface with an inoculating wire or loop. Large skin scales or whole nails

**Table 4.2.** Detection Times of Fungi Recovered by Isolator and Biphasic BHI Medium

| Species | Time to Detection (days) | | | | | | | | | |
|---|---|---|---|---|---|---|---|---|---|---|
| | Isolator | | | | | BHI | | | | |
| | Total Recovered | | | Matched Pairs | | Total Recovered | | | Matched Pairs | |
| | Mean | Median | Range | Mean | Median | Mean | Median | Range | Mean | Median |
| *C. albicans* | 2.37 | 2.0 | 1–6 | 1.71 | 2.0 | 4.0 | 3.0 | 2–8 | 3.14 | 3.0 |
| *C. glabrata* | 2.95 | 2.0 | 1–10 | 2.82 | 2.0 | 7.47 | 4.0 | 2–30 | 7.71 | 5.0 |
| *C. guilliermondii* | 1.0 | 1.0 | | 1.0 | 1.0 | 2.0 | 2.0 | 1–3 | 2.0 | 2.0 |
| *C. parapsilosis* | 2.0 | 2.0 | | 2.0 | 2.0 | 4.75 | 4.5 | 4–6 | 5.0 | 5.0 |
| *C. tropicalis* | 1.96 | 2.0 | 1–6 | 1.55 | 1.0 | 2.57 | 2.5 | 2–3 | 2.45 | 2.0 |
| *C. neoformans* | 2.25 | 2.0 | 2–3 | 2.0 | 2.0 | 4.5 | 4.5 | 4–5 | 4.0 | 4.0 |
| *Saccharomyces* sp | 3.0 | 3.0 | | | | | | | | |
| *T. beigelii* | 2.0 | 2.0 | | 2.0 | 2.0 | 2.5 | 2.5 | 2–3 | 3.0 | 3.0 |
| *Beauvaria* sp | 8.5 | 7.5 | 7–12 | | | | | | | |
| *H. capsulatum* | 9.39 | 8.0 | 7–14 | 8.0 | 8.0 | 15.67 | 23.0 | 15–40 | 24.14 | 23.0 |
| *N. asteroides*[a] | 14.0 | 14.0 | | | | | | | | |

[a] Acid-fast bacterium.

can be minced or ground prior to inoculating. Cultures can be incubated at room temperature (25° C) or at 30° C.

## CULTURING OF CLINICAL SPECIMENS

### Media

A number of satisfactory media are available for the primary recovery of fungi from clinical specimens. Table 4.1 lists the primary isolation media and the differential test media commonly used in clinical microbiology laboratories for the recovery and identification of fungi from specimens and their indications for use. Obviously chances for recovery can be increased by using a larger number of media; however, the use of only two or three are recommended. Table 4.3 presents common fungal culture media that can be recommended according to the type of clinical specimens. One may select a medium from each of the four categories shown in Table 4.3; however, at least one each from categories numbered 1, 2, and 3 should be used. It should also be mentioned that Sabouraud's dextrose agar, commonly used in many laboratories, is not satisfactory for the recovery of fungi from clinical specimens except for dermatophytes from skin scrapings or yeasts from vaginal cultures. Sabouraud's agar (2% dextrose) is, however, satisfactory for the subculture of fungi when sporulation studies are desired (see Table 4.4).

Specimens from sterile body sources (CSF) require fewer media than those that are contaminated with bacterial flora (respiratory tract). Specimens submitted for dermatophyte culture require only one medium as shown in Table 4.3. The formulations and methods for preparation of these media are included in Appendix II. The recovery of *Nocardia* sp from clinical specimens can be achieved by using processing and culturing techniques similar to those used to recover mycobacteria (72).

### Antibiotic Supplements

Since most clinical specimens submitted for fungal culture are collected from nonsterile sources, it is necessary to use media containing bacterial inhibitors to prevent the overgrowth of fungi by contaminating bacteria. A combination of gentamicin (5 μg/ml) and chloramphenicol (16 μg/ml) provides satisfactory inhibition of most bacteria. Since these antibiotics are heat stable, the media to which they are added can be autoclaved without reducing their antibacterial action.

In some instances, the rapidly growing molds may overgrow the slower growing organisms such as *H. capsulatum* and an antifungal compound, **cycloheximide (Actidione)** should be added to the culture medium. When this compound is added, the growth of many saprobic molds is inhibited; however, Actidione also inhibits the growth of many strains of *C. neoformans*, *Aspergillus* sp, *Candida* sp, *Pseudallescheria* (*Petriellidium*) *boydii*, and *Trichosporon beigelii*. One set of culture media not containing Actidione should always be included in the battery of media used.

Specimens referred to a laboratory by mail may be compromised by overgrowth with contaminating bacteria and rapidly growing saprobic molds. The use of a **yeast-extract phosphate medium** in combination with ammonium hydroxide has been reported to be helpful for the recovery of dimorphic pathogenic fungi from this type of specimen (82a).

### Blood Supplements

The addition of 5–10% sheep blood to fungal culture media can be recommended to enhance the recovery of some of the slower growing, more fastidious dimorphic fungi. Blood may be added to brain-heart

**Table 4.3.** Culture Media Useful for the Recovery of Fungi from Clinical Specimens

| | Culture Media | | | | | | | | | | |
|---|---|---|---|---|---|---|---|---|---|---|---|
| Specimens | Sabouraud's 2% dextrose | Mycosel (BBL)—Mycobiotic (Difco) | Yeast Extract Phosphate | Inhibitory Mold Agar | SABHI Agar | SABHI with C-G[a] | Brain-Heart Infusion (BHI) + C-G[a] | BHI Agar (Non-inhibitory) | BHI Agar + 10% sheep blood + C-G[b] | BHI Agar + C-G + Cy[c] | BHI Agar + C-G + Cy and 10% sheep blood[d] |
| Cutaneous: | | | | | | | | | | | |
| Skin | 4[f] | | 4 | 1 | 4 | 1 | 1 | 4 | 2 | 3 | 3 |
| Skin (dermatophyte) | | 1 | | | | | | | | | |
| Hair | | 1 | | | | | | | | | |
| Nails | | 1 | | | | | | | | | |
| Eye | 2 | | | 1 | 2 | | | 2 | 2 | 3 | 3 |
| Fluids: | | | | | | | | | | | |
| Cerebrospinal | 3 | | | 1 | 1 | | | 1 | 2 | 3 | 3 |
| Abdominal | 4 | | | 1 | 4 | 1 | 1 | 1 | 2 | 3 | 3 |
| Chest | 4 | | 4 | 1 | 4 | 1 | 1 | 1 | 2 | 3 | 3 |
| Synovial | 4 | | | 1 | 4 | 1 | 1 | 1 | 2 | 3 | 3 |
| Blood[e] | 3 | | | 1 | 4 | 1 | 1 | 1 | 2 | 3 | 3 |
| Bone marrow | 3 | | | 1 | 2 | | 1 | 2 | 2 | 3 | 3 |
| Gastric aspirates | 4 | | 4 | 1 | 4 | 1 | 1 | 4 | 2 | 3 | 3 |
| Genitourinary: | | | | | | | | | | | |
| Urine | 4 | | | 1 | 4 | 1 | 1 | 4 | 2 | 3 | 3 |
| Vaginal | 1 | | | 1 | | | | 1 | | | |
| Respiratory secretions | | | | | | | | | | | |
| Sputum | 4 | | 4 | 1 | 4 | 1 | 1 | 4 | 2 | 3 | 3 |
| Bronchial aspirates | 4 | | 4 | 1 | 4 | 1 | 1 | 4 | 2 | 3 | 3 |
| Transtracheal aspirates | 4 | | 4 | 1 | 4 | 1 | 1 | 4 | 2 | 3 | 3 |
| Mucous plugs | 4 | | 4 | 1 | 4 | 1 | 1 | 4 | 2 | 3 | 3 |
| Tracheal | 4 | | 4 | 1 | 4 | 1 | 1 | 4 | 2 | 3 | 3 |
| Ear | 4 | | 4 | 1 | 4 | 1 | 1 | 4 | 2 | 3 | 3 |
| Nose | 4 | | 4 | 1 | 4 | 1 | 1 | 4 | 2 | 3 | 3 |
| Nasopharynx | 4 | | 4 | 1 | 4 | 1 | 1 | 4 | 2 | 3 | 3 |
| Mouth | 4 | | 4 | 1 | 4 | 1 | 1 | 4 | 2 | 3 | 3 |
| Tissue and wounds | 4 | | 4 | 1 | 4 | 1 | 1 | 4 | 2 | 3 | 3 |

[a] Contains chloramphenicol, 16 μg/ml, and gentamicin, 5 μg/ml.
[b] Contains chloramphenicol, 16 μg/ml, gentamicin 5 μg/ml, and 10% sheep blood.
[c] Contains chloramphenicol, 16 μg/ml, gentamicin 5 μg/ml, and cycloheximide 500 μg/ml.
[d] Contains chloramphenicol 16 μg/ml, gentamicin 5 μg/ml, cycloheximide 500 μg/ml, and 10% sheep blood.
[e] Used in combination with Dupont Isolator.
[f] Listed in order of importance. Note alternate media for the same number; choose one medium from at least 3 of the 4 categories listed for each culture source.

**Table 4.4.** Media Recommended for Subculture of Filamentous Fungi

| Fungal Group | Media[a] |
|---|---|
| Zygomycetes | Cornmeal agar<br>Inhibitory mold agar<br>Sabouraud's dextrose agar (2%) |
| Hyaline Molds | Inhibitory mold agar<br>Cornmeal agar<br>Sabouraud's dextrose agar (2%) |
| *Aspergillus* sp | Czapek Dox agar |
| Dermatophytes | Cornmeal agar |
| Dimorphic molds | Inhibitory mold agar<br>Yeast extract agar |
| Dematiaceous molds | Cornmeal agar<br>Lactrimel |

[a] Formulae for the media listed in this table can be found in the Appendix II.

infusion-broth that may be inhibitory to the growth of some pathogenic fungi.

## CULTURE DISHES VS. TUBES

Each laboratory director must decide whether to use culture dishes or tubes for the primary recovery of fungi from clinical specimens. Large culture tubes are recommended for those laboratories that may be either unfamiliar with the proper handling of culture plates or that do not have adequate biological safety hoods. The culture tubes should be fitted with cotton plugs or screw-cap tops. If caps are used, they should be slightly loosened after inoculation to allow the cultures to "breathe." The media should be poured in thick slants to prevent dehydration during storage.

The disadvantage of culture tubes is that the surface area is limited for the satisfactory isolation of colonies. Culture dishes, in contrast, provide a large surface on which mixed cultures can be more readily observed, and colonies have maximal aeration. The use of tubes has an advantage in laboratories where space is a problem as they take up little room in storage or during incubation.

The tendency for media in dishes to dry out during storage or prolonged incubation can be minimized by pouring at least 40 ml of agar into each dish or sealing them if necessary in oxygen-permeable cellophane bags or with oxygen-permeable tape. Dishes should be taped on either side to prevent inadvertent opening during incubation.

The potential for contaminating the environment or contracting laboratory-acquired infection is greater when dishes are used. It is imperative when working with dishes that they be opened and examined only within an adequately vented biological safety hood.

### Incubation of Cultures

All fungal cultures are incubated at room temperature or preferably at 30° C and held for 30 days before discarding as negative. Although some fungi require incubation at 35°–37° C to demonstrate their yeast forms, primary recovery from clinical specimens should be done at 30°C. Molds that grow up can sometimes be converted to their yeast forms by subculturing a portion of the colony to appropriate conversion media and incubating at 35°–37° C. A relative humidity of 40–50% can be maintained in the incubators by placing an open pan of water under the bottom shelf or by using a humidified incubator.

### Subcultures of Fungal Colonies

It is often necessary to transfer fungal colonies from primary recovery media to secondary media, particularly if the primary plate shows more than one fungal species or is heavily contaminated with bacteria. Subculture to differential testing media may also be required to study specific growth characteristics or to confirm biochemically an unknown species if the identification is not obvious from study of the colonial or microscopic morphology alone. Table 4.4 lists the media that should be considered for the subculture of different groups of filamentous molds.

In making subcultures, a small portion of the primary colony should be picked with a sterile, bent inoculating wire and transferred to the surface of the secondary plate. The agar of the secondary plate should be stabbed with the transfer wire so that a portion of the colony being transferred is placed beneath the agar surface. One should obtain the subculture from an area on the primary colony where sporulation is active. A point midway between the center and the periphery of those colonies with an entire margin is usually optimal; for colonies that grow to the rim of the Petri dish without a margin, sampling from the most granular portion is recommended. Subcultures should also be incubated at 30° C, unless studies at different temperatures are required. Cultures should be examined daily since growth in subculture is usually more rapid than the time required for primary isolation.

The procedures to make an identification of fungal colonies both in primary isolation and in subculture are discussed in detail in the next chapter.

# Chapter 5

# PRELIMINARY IDENTIFICATION OF FUNGAL CULTURES

## READING OF CULTURES

Ideally, examination of cultures for growth should be made daily, at least during the first week or two of incubation; however, examinations three times per week is also satisfactory. Cultures may be examined on a twice weekly schedule thereafter. Since many fungal colonies grow rapidly, their appearance on the surface of agar media usually can be detected by the naked eye. Observations can be made by holding and tipping the plate back and forth with one hand to reflect the incident light at an appropriate angle. A hand lens or dissecting microscope may be helpful in the early detection of microscopic growth, but such devices are not usually necessary.

Biphasic blood culture bottles are examined with transmitted light from an incandescent lamp either for turbidity and/or flecks of growth in the broth or for the presence of colonies on the agar slant. All negative cultures should be gently agitated daily so that the blood-broth mixture is gently flooded over the agar surface. Cerebrospinal fluid cultures prepared with the microfilter technique should be examined by gently lifting the filter paper pad observing for growth in the area under the pad. The pad is then placed in a new location every other day during the first week of incubation, after which time it can be left in place and observed for growth around the disk margins.

It is generally possible to determine from visual examination whether a fungal colony is a mold or a yeast, an important assessment for the initial categorization of isolates (see Table 5.3). Mold colonies have a fuzzy or stringy appearance from the growth of aerial hyphae (Color Plate 2); yeasts, however, produce buttery or pasty colonies with a smooth surface, sometimes simulating bacterial colonies. Observations should be made for the consistency of the surface growth, for the patterns of folding (rugae), for the distinctness of the colony margin and for the presence of pigment either on the surface or the reverse of the colony or diffusing into the surrounding medium. Both the surface and the back side of the colony should be examined routinely.

For the examination of molds, the use of a dissecting microscope may be helpful in determining the pattern of mycelial growth and in detecting the presence and morphological characteristics of reproductive fruiting bodies. Although these observations may be somewhat helpful, the assessment of colonial morphology generally is of limited value in identifying fungi because of variations caused by differences in environmental conditions during incubation and in the types of culture media used. The same species of fungi may not appear alike when cultured on different days; or, different species of fungi may appear similar. Therefore, the preparation of slide mounts from small portions of the colonies for microscopic study is usually necessary before a final report can be issued.

## PREPARATIONS FOR MICROSCOPIC STUDY OF FUNGI

Several methods can be used for the microscopic examination of fungi. In each instance, it is necessary to transfer a small portion of the colony to a glass slide in a drop or two of water or suitable mounting medium. Since the refractive indices of the microscopic structures of many fungi is near that of water, either phase contrast microscopy or a staining technique must be used for adequate study. The mounting/staining medium most commonly used is lactophenol aniline blue (lactophenol cotton blue).

### Wet Mount Method

The use of wet mounts is a time-honored and rapid method for preparing fungal colonies for microscopic examination. The procedure is performed by taking a bent needle, made of heavy gauge wire, and removing a small portion of the colony to be studied, to include a small portion of the underlying agar. The sample should be taken at a place intermediate between the center and periphery of the colony. This small colony fragment and supporting agar is transferred to a drop of lactophenol-aniline blue on a glass microscope slide and a cover slip put in place. Pressure is applied directly over the colony fragment using a pencil eraser to depress the hyphae and other structures to enhance their microscopic examination (Fig. 5.1). The lactophenol-aniline blue solution serves not only as a stain but the phenol also serves to kill the portion of culture under examination to minimize the chances for laboratory-acquired infections.

The wet mount can be easily and quickly prepared and often is sufficient to make an identification for many of the fungi encountered in the laboratory. The major drawback of the wet mount is the difficulty in preserving continuity between the spores, fruiting structures, and hyphae because of the rigorous treatment. This fault may be critical when it is necessary to make an exact microscopic definition of some of the fungal species with delicate structures. The **Scotch tape technique** is commonly employed to overcome this problem.

### Scotch Tape Technique

Scotch brand tape, No. 600, or equivalent, is recommended because it is highly transparent. Obviously frosted tapes will not be suitable for this technique. A 4-cm strip is looped back on itself, with the adhesive side out, and held between the thumb and index finger (Fig. 5.2). The adhesive side is then pressed firmly to the surface of the fungal colony to be studied. The aerial hyphae cling to the sticky surface and can be gently pulled from the mat. Obviously this technique will be of limited value in the study of colonies with a smooth or yeast-like surface.

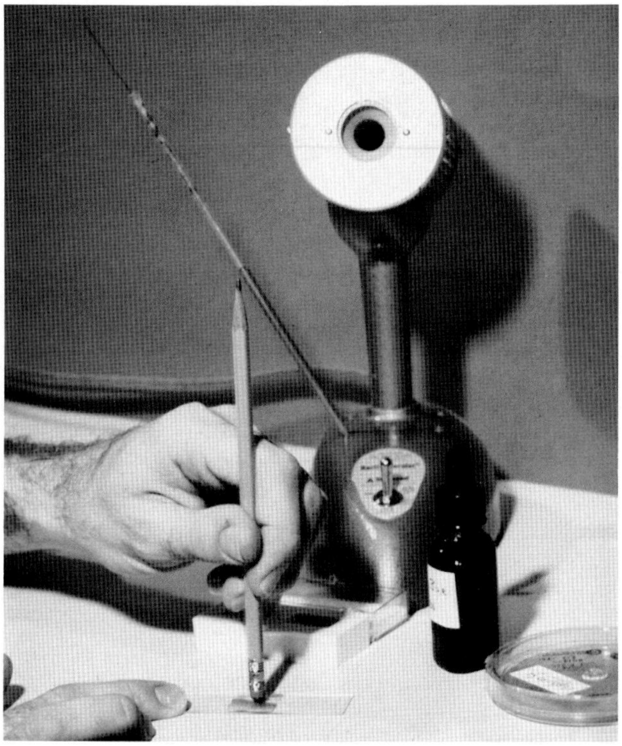

**Figure 5.1.** Step in the wet mount procedure illustrating the use of a pencil eraser to gently press the cover slip over the specimen/lactophenol-aniline blue mount.

The inoculated tape strip is placed in a small drop of lactophenol aniline blue mounting fluid on a microscope slide (Fig. 5.3). This preparation is usually superior to tease mounts because the original juxtaposition of the spores and hyphal segments is retained.

If the tape is not pressed firmly enough to the colony surface, the sample may not be adequate. In instances where sporulation is not present, a wet mount should

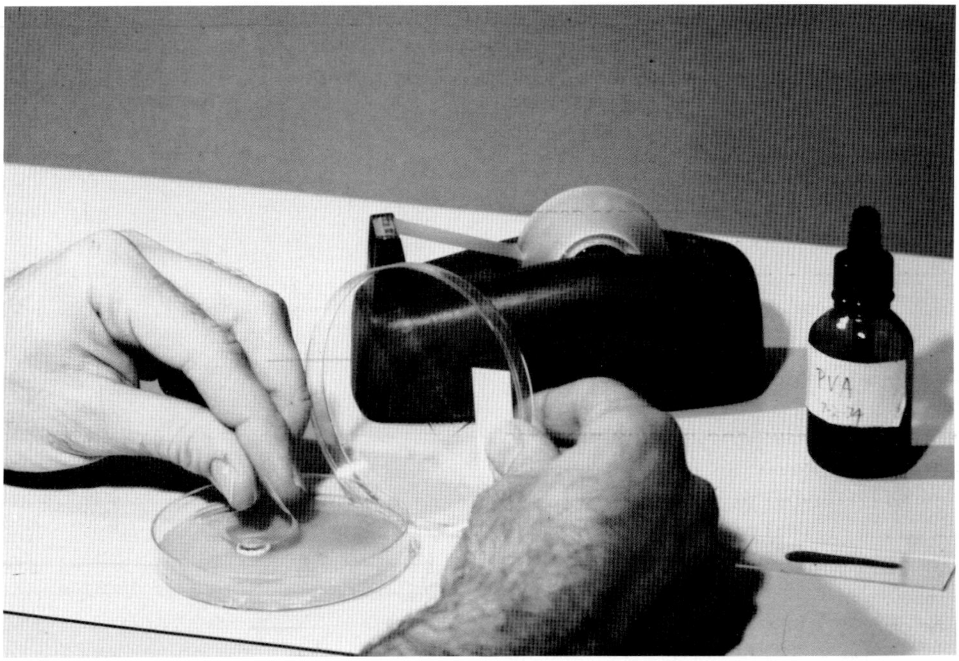


**Figure 5.2.** Step in the Scotch tape procedure illustrating sampling of a fungus colony with the adhesive side of the tape.

be made. We have observed the macroconidia of *Histoplasma capsulatum* in wet mount preparations where the Scotch tape preparations revealed only hyphal fragments. Conversely, there are instances where cultures that sporulate heavily reveal only conidia when Scotch tape preparations are observed. A second mount should be made from the periphery of the colony where growth and sporulation is not as heavy.

Scotch tape preparations are difficult to make from fungal colonies growing in narrow-mouth tubed media. If such studies are necessary, each end of a 2-cm long piece of tape can be stuck to the ends of applicator sticks which can then be rolled outward to produce a small cylinder of tape with the sticky side out. These sticks can then be inserted into the neck of the tube and touched to the surface of the colony, taking care not to touch the sides of the glass with the Scotch tape on the way in or out. If it is not possible to prepare a Scotch tape mount for the reasons cited above; or, if the microscopic structures still are not adequately preserved for study, a microculture should be prepared.

### Microculture Method

The preparation of the microculture (also known as slide culture) is illustrated by the photographs included in Plate 5.1. Although this method is somewhat cum-

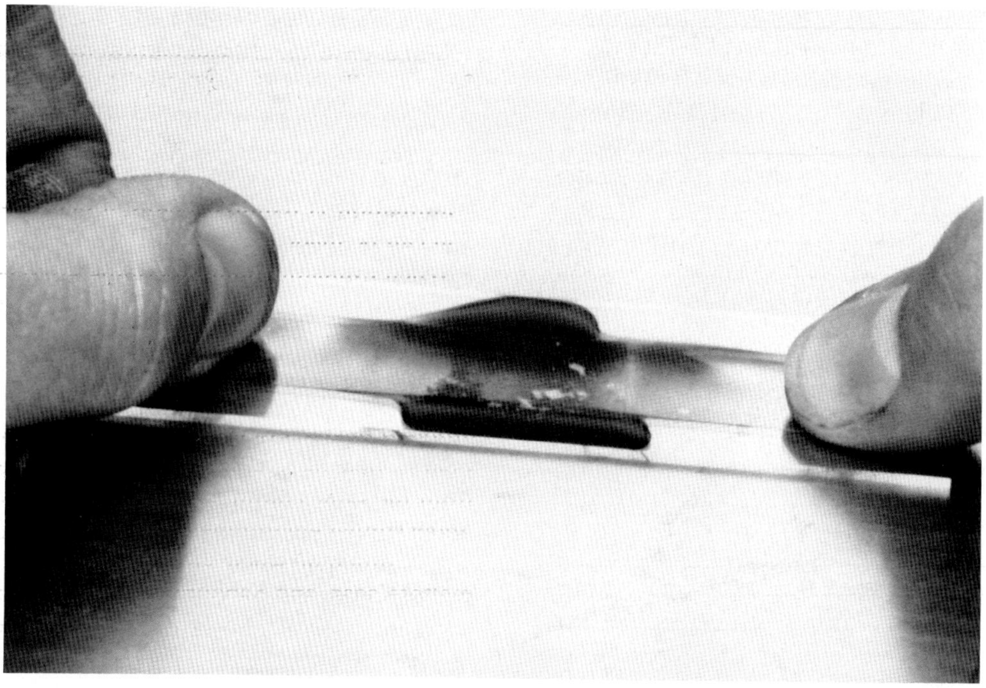

**Figure 5.3.** Step in the Scotch tape procedure illustrating how the inoculated tape is affixed over a drop of lactophenol aniline blue on a microscope slide.

bersome to perform and poorly suited for making a rapid diagnosis, it is unparalleled for demonstrating the subtle microscopic features necessary to make the definitive identification of some fungi, or for preparing semipermanent mounts to be used for teaching sets. The procedure is carried out as follows:

**Slide Culture Method (Plate 5.1)**

1. Cut a small block of potato dextrose or cornmeal agar that has been previously poured in a Petri dish to a depth of 4 mm. This may be done by using a sterile surgical blade or sharp-edged spatula or a sterile 18 × 150 mm straight bore test tube having no lip (Plate 5.1*A*).
2. On the agar surface of a second Petri dish, place a sterile microscope slide (Plate 5.1*B*). Alternatively, place a round piece of filter paper and two applicator sticks, cut at such a size to fit into the dish. Place a sterile 3 × 1 cm glass microscope slide on top of the applicator sticks (Plate 5.1*C*).

3. With a sterile bent wire, place the agar block to the surface of the microscope slide (Plate 5.1*D*).
4. With the sterile wire, remove small portions of the fungal colony to be studied and inoculate the four quadrants of the agar block (Plate 5.1*E*).
5. After inoculation, place a sterile number 1 cover slip on the agar surface (Plate 5.1*F*).
6. Saturate the filter paper in the bottom of the Petri dish (if the alternate method was used). Replace the lid and incubate the mount at 30° C.
7. The colony will grow under the surface of the cover slip (Plate 5.1*G*). Visually examine the mount periodically to determine when the colony has matured and is ready for harvest.
8. When sufficient growth is evident, carefully remove the coverslip with sterile forceps and place it on a microscope slide containing one drop of lactophenol-aniline blue mounting fluid (Plate 5.1*H*). If a permanent mount is desired, wipe clean the surface of the glass slide immediately adjacent to the edge of the cover slip and generously apply clear fingernail polish or mounting fluid.
9. The original slide will also serve as a second mount. Gently remove the agar block from the original slide and decant it into a beaker or plastic cup containing a solution of 5% phenol.
10. Apply one drop of lactophenol-aniline blue solution to the area where the block had been in contact and overlay with a number 1 cover slip. This preparation can also be ringed with clear fingernail polish or mounting fluid if a semipermanent preparation is desired.

### Gross Appearance of Fungal Colonies

Fungi grow on artificial culture media in a variety of forms as illustrated in Plate 5.2. Although it may be possible to recognize a well-known species recovered repeatedly in a given geographic area based on its gross colonial morphology alone, most often microscopic examination is required before a definitive identification can be made.

When evaluating and describing gross colonial morphology, one must consider the type of culture media employed and the environmental conditions under which the fungus was grown. The gross appearance of the same fungus may vary greatly when cultured on different media. For example, note in Figure 1 of Color Plate 2 that the colonies of *A. glaucus*, all inoculated from the same culture at the same time and incubated under similar environmental conditions, appear considerably different when grown on Czapek's agar, 20% maltose agar, and Sabouraud's dextrose agar. Another example of how colonial appearance is media dependent is the 30° C form of the dimorphic fungus, *H. capsulatum* which appears as a mold on brain-heart infusion agar, but appears yeast-like in appearance when grown on the same media containing 5% sheep blood.

A variety of terms may be used to describe the gross appearance of fungal colonies. For example, colonies that present a well-defined aerial mycelium may be described as fluffy, cottony, wooly, or cobweb in appearance depending upon the terminology preferred by the observer and the exact consistency of the culture (Plate 5.2*A* and *B* and Color Plate 2). Colonies that have a well-developed aerial mycelium may appear fuzzy or hairy; those in which the aerial mycelium is very poorly developed or is absent altogether produce colonies with a smooth, or glabrous surface (Plate 5.2*C* and Color Plate 5). Yeast colonies in particular are good examples of a glabrous surface, and the encapsulated strains of *C. neoformans* may appear shiny, mucoid, and slimy (Plate 5.2*D*). Small glabrous colonies may also have a brittle, chalky consistency, suggestive of *Nocardia* or *Streptomyces* sp.

If the aerial mycelium has a delicate silky appearance, one should be suspicious of one of the dimorphic pathogens and take appropriate precautions in handling the culture (Color Plate 5). Dimorphic fungi may also produce spiked or prickly colonies, particularly at 35° C incubation during the conversion of mold to yeast forms (Plate 5.2*E* and Figure 3 of Color Plate 5).

Fungi that sporulate heavily have a granular or sugary colonial appearance (Plate 5.2*F* and Color Plates 2–4). The granular portion often is deeply pigmented, producing a wide variety of colors, as shown in Color Plate 2. The white, fluffy apron at the periphery of a colony represents the growing outer margin, too young for spores to have formed (Figure 1 of Color Plate 4). It is in the older, central portions of the cultures where sporulation is heaviest and intensity of surface pigmentation is greatest. Fungal colonies also often develop irregular folds or "rugae." Rugae often eminate radially from the center of the colony in a spoke-wheel fashion (Plate 5.2*F*, Figures 2 and 3 of Color plate 2 and Figure 5 of Color Plate 3).

Zygomycetes produce rapidly growing colonies, initially white but turning brown or gray-black with age, that extend from one border of the Petri dish to the other, filling the dish with a dense woolly mycelium (Plate 5.2*G* and Figures 4 and 5 of Color Plate 4). This type of growth is virtually diagnostic of Zygomycetes.

Colonies may also develop a rapidly appearing tuft of fluffy hyphae usually in the center of the colonies (Figure 6 of Color Plate 3). These umbonate formations often are the initial indication that a colony may be turning sterile (absence of reproductive spore formation) and should be avoided when subculturing to other media.

Colonies with a brown or black pigmentation are referred to as dematiaceous, meaning dark (Plate 5.2*H* and Color Plate 1). Dematiaceous species of fungi are dark because the vegetative hyphae and or spores are pigmented; therefore, the reverse of the colony also appears black. It is important to observe the reverse side of a colony for pigmentation in making a preliminary identification of the dematiaceous group of fungi, to differentiate them from *A. niger* (Figure 3 of Color Plate 4), for example, which may appear black on the surface due to the dense production of dark spores, but will appear light tan or buff on the reverse; true dematiaceous fungi produce black colonies even when quite young and small.

All fungal cultures should be handled with care in the laboratory, particularly if the growth is relatively slow (5 days or more) and has a silky or cobweb appearance (Figures 2, 4, and 6 of Color Plate 5). These are features indicative of one of the dimorphic pathogens with high potential for causing laboratory acquired infections. If one is suspicious that a dimorphic pathogen has been recovered, a small amount of sterile water can be used to wet down the surface of the colony before preparing microscopic mounts. All fungal colonies should be examined and handled only under a properly operating microbiological safety hood.

### Microscopic Description of Fungal Colonies

As discussed above, microscopic examination of fungal cultures is usually necessary to establish a presumptive or definitive identification. In contrast to the laboratory work-up of most bacteria in which a battery of biochemical tests can be performed to make a definitive identification, very few such tests are available to identify the fungi and one must rely almost totally on the recognition of certain microscopic morphological features. The microscopic characteristics by which the medically significant fungi can be identified in culture will be presented in the following chapters; here, we wish to present only some of the basic structural forms that are common to many of the fungi and may be observed in a variety of clinical isolates. As with gross colony descriptions, mycologists have devised specific terms to refer to these various structures, examples of which are illustrated in Plate 5.3.

### Basic Structural Forms

The fundamental structural units of fungi are the tube-like projections known as hyphae. If hyphae are not divided by cross walls, they are **aseptate** (Plate 5.3*A*); if cross walls are present, they are **septate** (Plate 5.3*B*). Because the protoplasm within the aseptate hyphae can run along the length of the strand uninterrupted by septa, these hyphae are also spoken of as "**coenocytic.**"

A number of hyphae intertwine to form a loose mat called the **mycelium.** In practice, the terms hyphae and mycelium are used interchangeably and it is not uncommon to hear the terms aseptate mycelium and septate mycelium.

The nutrient-absorbing and water-exchanging portion of a fungus, extending downward into the substrate, is called the **vegetative mycelium.** The portion extending above the surface is known as the **aerial mycelium;** or, alternatively as the reproductive mycelium since they often support fruiting bodies from which spores or conidia are often derived.

## FORMS OF VEGETATIVE HYPHAE

Vegetative hyphae may form a variety of nonspecific structures that have little diagnostic significance since they may be seen in a variety of fungal species. The dermatophyte group of fungi in particular have a predelection for producing these forms:

**Forms of Vegetative Hyphae**

| | |
|---|---|
| **Racquet hyphae:** | Club-shaped segments with the smaller ends adjoining the broad ends of an adjacent one (Plate 5.3*C*) |
| **Favic chandeliers:** | Broad, terminal hyphal branches with rounded or blunt ends simulating the antlers of a buck deer (Plate 5.3*D*) |
| **Pectinate bodies:** | Parallel extensions projecting at right angles to the hyphal strand, simulating the teeth of a comb, thus, the term "pectinate" (Plate 5.3*E*) |
| **Spiral hyphae:** | Corkscrew-like turns of mycelium which are commonly seen in older fluffy cultures of *Trichophyton* sp (Plate 5.3*F*) |

## SPORULATION

Sporulation of fungi may be either asexual or sexual depending upon the manner in which reproductive spores are formed. Species of fungi which are capable of sexual reproduction are known as **perfect fungi**; those in which sexual sporulation has not been demonstrated comprise the **fungi imperfecti**.

***Sexual Sporulation***. Sexual sporulation involves the formation of two separately developed nuclei which fuse followed by meiosis resulting in a sexual spore.

Some species of perfect fungi produce sexual spores in a large closed, bag-like structure called a **cleistothecium** (Plate 5.3*G*). The cleistothecium, in turn, contains smaller sacs called **asci**, each of which contains four or eight **ascospores** (Plate 5.3*H*).

Another bag-like structure, closely resembling a cleistothecium, that is asexually produced by imperfect fungi is called a **pycnidium** (Plate 5.3*I*). Pycnidia contain conidia, which can be distinguished from ascospores by their smaller size (Plate 5.3*I*). The contents of cleistothecia or pycnidia can be observed in wet preparations or lactophenol-aniline blue mounts by gently depressing the surface of the cover slip with the eraser of a pencil while looking under the microscope and watching for the bag-like structures to rupture under the increased pressure. Cleistothecia will exude ascospores; pycnidia will exude conidia (Plate 5.3*I*).

Most of the fungi of medical importance belong to the **fungi imperfecti**. There is a general feeling among mycologists that all fungi probably possess a sexual stage, although the perfect form has not been identified in artificial culture media.

***Asexual Sporulation***. Asexual spores may be derived either from the vegetative mycelium or may develop from specialized supporting hyphae which project above the surface of the substrate.

***Conidia Derived from the Vegetative Mycelium***. Three types of spores are derived directly from the vegetative mycelium:

**Vegetative Conidia**

| | |
|---|---|
| **Arthroconidia:** | Formed directly from the hyphae by fragmentation through points of septation. Where mature, they appear as square, rectangular, or barrel-shaped, thick-walled cells (Plate 5.3*K*). When stained with lactophenol aniline blue, some species stain regularly; other species, notably the dimorphic pathogen, *Coccidioides immitis*, show alternate staining arthroconidia separated by empty, unstained segments. |
| **Blastoconidia:** | Produced by budding, with the daughter cells pinching off from portions of the mother cells through sausage-like constrictions (Plate 5.3*L*). This is the typical type of sporulation found in budding yeasts. In some species of yeasts, notably *Candida* sp, blastoconidia may elongate into chain-like formations known as **pseudohyphae** (Plate 5.3*M*). |
| **Chlamydospores:** | Round, thick-walled, differentiated areas formed directly from the hyphae. Those remaining within the hyphal strand are called **intercalary**; those forming to the side are **sessile**; those at the end of the strand are **terminal** (Plate 5.3*N*, *O*, and *P*). |

## AERIAL SPORULATION

Aerial sporulation is more elaborate than vegetative sporulation, involving the generation of specialized supporting hyphae that project above the surface of the substrate. The suffix "phore" (carrying) is appended to several prefixes in referring to these specialized supporting hyphae (sporangiophore, conidiophore, etc). Aerial sporulation takes on several morphological forms that are helpful in establishing a definitive identification.

## SPORANGIA

A **sporangium** is a large, sac-like structure that is supported by a specialized, long, slender, at times branching sporophore called a **sporangiophore** (Plate 5.4*A*). The spores formed within these enclosures are **sporangiospores.** This type of aerial sporulation is characteristic of the group of fungi belonging to the Zygomycetes.

## CONIDIA

Most other groups of fungi with an aerial mycelium bear spores freely from the surface of a variety of fruiting bodies. These spores are referred to as "**conidia,**" derived from a Greek term meaning "dust." Appropriately, the specialized supporting sporophores are called **conidiophores**. Surface-bearing spores that break away from the main hyphal strand by the fracture of a connecting cell are called **aleuriospores**.

A variety of conidial sporulations is shown in Plate 5.4. Plate 5.4*B* and *C* illustrates conidiophores that terminate in a swollen structure called a **vesicle**. From the surface of the vesicle are formed secondary extensions known as **phialides**, which in turn give rise to long chains of conidia (Plate 5.4*C*). This type of fruiting structure is characteristic of *Aspergillus* sp.

Plate 5.4*D* illustrates a conidiophore that freely branches into a structure known as a **penicillus**, with each branch terminating in phialides from which chains of spores are derived. *Penicillium*, *Paeciliomyces*, and *Scopulariopsis* sp, to be described in Chapter 6, are prototype fungi that sporulate from this type of a penicillus structure.

A single, slender conidiophore (phialide) that supports a fruiting head in which the conidia are arranged in clumps, held together within a mucinous sheath that prevents the conidia from becoming dispersed, is illustrated in Plate 5.4*E*. An example of conidia arranging in clumps is *Acremonium* (*Cephalosporium*) sp.

Conidia that are quite small and singly attached to the hyphal strands via delicate connecting cells are called **microconidia** (Plate 5.4*F*). They can be round (spherical), egg-shaped or oval (elliptical), pear-shaped (pyriform), or club-shaped (clavate).

**Macroconidia,** a term referring to larger, thick-walled, usually multicelled conidia, are illustrated in Plate 5.4*G*. Macroconidia are generally spindle-shaped (fusiform) or clavate and may have smooth or roughened walls. Microconidia and macroconidia are seen in many fungal species and are not specific, except as they are used to differentiate the genera of the dermatophytes (described in Chapter 6). In some instances the conidiophores of some organisms are observed in tightly bend, upright bundles called **coremia** (Plate 5.4*H*).

## SYSTEMATIC MYCOLOGY

The microscopic structures described above are most frequently used by clinical mycologists in the identification of fungal isolates. However, the recognition of several other microscopic structures and the manner in which conidia are formed from the conidia-bearing portion of the mycelium (a process called conidiation) is of importance to systematists and taxonomists, particularly for the initial identification and classification of fungi. Many current textbooks present a major section on conidiogenesis; however, electron microscopy is an integral part of the study of the origin of conidia production and is not within the scope of a clinical microbiology laboratory. Therefore, we have not included a section on conidiogenesis in this text and readers are referred to current texts in general mycology, notably by Rippon (70) and by McGinnis (Suggested Readings) for this information.

## TAXONOMY

In the preliminary identification of fungal isolates from clinical materials, it is helpful to have in mind a scheme of classification. Several classifications have been proposed, none of which are suitable for all purposes. The term fungus in general usage refers to molds and yeasts; technically they are organisms within a separate kingdom characterized by the lack of true leaves, stems, or roots. They lack chlorophyll and therefore cannot derive energy from photosynthesis, rather derive their nutrition through absorption of preformed organic compounds. Morphologically they are unicellular or mycelial.

The phylogenetic classification has not been of practical use in the clinical laboratory in the identification of fungi. By this classification, the pathogenic fungi can be included within four phyla: I—*Zygomycota*; II—*Ascomycota*; III—*Basidiomycota*; and IV—*Deuteromycota*. The **Zygomycetes** possess aseptate hyphae and reproduce asexually by means of spores borne in sporangia. Of the genera listed in Table 5.1, *Mucor* and *Rhizopus* are most frequently recovered from clinical specimens.

The **Ascomycetes** have a septate mycelium and reproduce either asexually via conidia borne on a wide variety of conidiophores or sexually by means of ascospores produced within a bag-like structure called an ascus. Many of the genera listed in Table 5.1, such as *Endomyces*, *Sortorya*, *Ajellomyces*, *Arthroderma*, and *Nannizzia*, are the perfect forms of various species of fungi that are rarely recovered in this form; rather, grow on artificial laboratory culture media in their imperfect, conidia-bearing forms. *P. boydii* and *A. nidulans* are exceptions where the perfect forms may be recovered in the laboratory and the ascospore-bearing cleistothecia or perithecia may be observed in microscopic preparations.

The **Basidiomycetes** are septate and produce either asexually by producing conidia or sexually by means of basidiospores. *Filobasidiella,* the perfect form of *C. neoformans*, is the only species within this phylum that may be seen in clinical cultures.

Most fungal isolates recovered in the clinical laboratory belong to the phylum **Deuteromycota**, as can be seen by the long list of species in Table 5.1. The **Deuteromycetes** have a septate mycelium and reproduce virtually exclusively asexually through the production of conidia. A detailed description of the genera of fungi listed under the Deuteromycota in Table 5.1 can be found in Chapter 6.

Clinicians, who must treat patients with local or disseminated mycoses, have found a disease-oriented classification of the fungi more useful (see Table 5.2). In this scheme, the fungi are grouped into three broad categories: (a) the superficial or cutaneous mycoses; (b) the subcutaneous mycoses; and (c) the systemic mycoses.

**Superficial or cutaneous mycoses** are fungal diseases that remain confined to the outer layers of the skin, nails, or hair without invading the deeper tissue or viscera. These mycoses are most commonly caused by one of the three groups of dermatophytes, *Microsporum*, *Trichophyton*, and *Epidermophyton*. The very superficial mycoses, tinea versicolor and piedra are confined to the outer keratinized portion of the skin. *Can-*

**Table 5.1.** Phylogenetic Position of Medically Significant Fungi[a]

| Class | Order | Family | Genus/Species |
|---|---|---|---|
| **Phylum Zygomycota** | | | |
| Zygomycetes | Entomophthorales | Entomophthoraceae | Basidiobolus |
| | Mucorales | Mucoraceae | Absidia |
| | | | Cunninghamella |
| | | | Mucor |
| | | | Rhizopus |
| | | | Syncephalastrum |
| **Phylum Ascomycota** | | | |
| Hemiascomycetes | Endomycetales | Endomycetaceae | **Endomyces** (*Geotrichum* sp)[b] |
| | | Saccharomycetaceae | **Kluyveromyces** (*Candida pseudotropicalis*)[b] |
| Loculoascomycetes | Myriangiales | Saccardinulaceae | *Piedraia hortae* |
| | Microascales | Microascaceae | *Pseudallescheria boydii* |
| Plectomycetes | Eurotiales | Eurotiaceae | **Emericella** (*Aspergillus nidulans*)[b] |
| | | | **Sortorya** (*Aspergillus fumigatus*)[b] |
| | | Gymnoascaceae | **Ajellomyces** (*Histoplasma capsulatum*)[b] |
| | | | **Arthroderma** (*Trichophyton* species)[b] |
| | | | **Nannizzia** (*Microsporum* species)[b] |
| **Phylum Basidiomycota** | | | |
| Teliomycetes | Ustilaginales | Filobasidiaceae | **Filobasidiella** (*Cryptococcus neoformans*)[b] |
| **Form Phylum Deuteromycota** | | | |
| Blastomycetes | | Cryptococcaceae | Candida |
| Hyphomycetes | Moniliales | Moniliaceae | Acremonium |
| | | | Aspergillus |
| | | | Blastomyces |
| | | | Chrysosporium |

**Table 5.1.** (continued)

| Class | Order | Family | Genus/Species |
|---|---|---|---|
| | | | Coccidioides |
| | | | Epidermophyton |
| | | | Geotrichum |
| | | | Gliocladium |
| | | | Histoplasma |
| | | | Microsporum |
| | | | Paecilomyces |
| | | | Paracoccidioides |
| | | | Penicillium |
| | | | Sepedonium |
| | | | Scopulariopsis |
| | | | Sporothrix |
| | | | Trichoderma |
| | | | Trichophyton |
| | | Dematiaceae | Alternaria |
| | | | Aureobasidium |
| | | | Cladosporium |
| | | | Curvularia |
| | | | Drechslera |
| | | | Exophiala |
| | | | Fonsecaea |
| | | | Helminthosporium |
| | | | Madurella |
| | | | Nigrospora |
| | | | Phialophora |
| | | | Rhinocladiella |
| | | | Stemphylium |
| | | | Ulocladium |
| | | | Wangiella |
| | | Tuberculariaceae | Epicoccum |
| | | | Fusarium |
| Ceolomycetes | Sphaeropsidales | | Phoma |

[a] Modified from Chandler FW, Kaplan W, and Ajello L, *Histopathology of Mycotic Disease*, Chicago, Year Book Medical Publishers, 1980.
[b] Genus/species designations for the asexual, imperfect forms are shown in parentheses.

*dida* sp can also closely simulate the dermatomycoses clinically and cultures must be obtained to make the differentiation.

**Subcutaneous mycoses** are confined to the deep subcutaneous tissue, with only rare reports of systemic spread. The dimorphic fungus *Sporothrix schenckii* is

**Table 5.2.** Disease-oriented Taxonomy of Pathogenic Fungi

| Cutaneous | Subcutaneous | Systemic |
|---|---|---|
| Superficial Mycoses | Chromoblastomycosis | Aspergillosis |
| Tinea | Cladosporium | Actinomycosis[a] |
| Piedra | Fonsecaea | Blastomycosis |
| Candidosis | Phialophora | Candidosis |
| Dermatophytosis | Sporotrichosis | Coccidioidomycosis |
| Microsporum | Phaeohyphomycosis | Cryptococcosis |
| Epidermophyton | Exophiala | Geotrichosis |
| Trichophyton | Phialophora | Histoplasmosis |
| | Wangiella | Nocardiosis[a] |
| | Mycetoma | Paracoccidiomycosis |
| | Actinomycotic | Saprobis sp. (rare) |
| | Actinonomycosis[a] | Sporotrichosis (rarely systemic) |
| | Nocardiosis[a] | Zygomycosis |
| | Eumycotic | |
| | Exophiala | |
| | Pseudallescheria | |

[a] These diseases are caused by filamentous bacteria but are commonly included with the fungi.

the most common cause of subcutaneous mycoses in the United States and is characterized by primary subcutaneous ulcers of the arms or legs with secondary ulcers forming along the tracts of the lymphatic channels. Less commonly encountered are two types of subcutaneous disease which are more prevalent in the tropical regions of the world: (a) **Mycetomas**, or granulomatous tumors of the subcutaneous tissue caused by a wide variety of mycotic and bacterial agents can be divided into two major groups; **eumycotic mycetomas** in which the etiological agents are true fungi producing hyphae and spore forms and **actinomycotic mycetomas**, caused by a group of filamentous bacteria that are commonly discussed with the fungi which they clinically and culturally simulate. (b) **Chromoblastomycosis** is characterized clinically by exophytic, verrucous, papillary granulomatous lesions caused by a variety of black (dematiaceous) molds to be discussed in more detail in Chapter 6.

Systemic mycoses may involve the deep viscera and become widely disseminated throughout the body. These mycoses are caused by fungi most commonly belonging to the genera *Blastomyces*, *Histoplasma*, *Paracoccidioides*, and *Coccidioides*, although many other species of fungi can also become disseminated, particularly in immunosuppressed hosts.

Table 5.2 lists the genera of fungi that are of medical importance within these three broad groups.

### Laboratory Classification: A Practical Working Schema

The phylogenetic and the medically oriented taxonomies have been of limited usefulness to medically trained technologists and microbiologists who work in clinical microbiology laboratories. The reason is one of perspective and not that any one taxonomic schema is more correct than the other. It is perhaps unfortunate that the training of medical technologists does not

**Table 5.3.** Commonly Encountered Fungi of Clinical Laboratory Importance: A Practical Working Schema

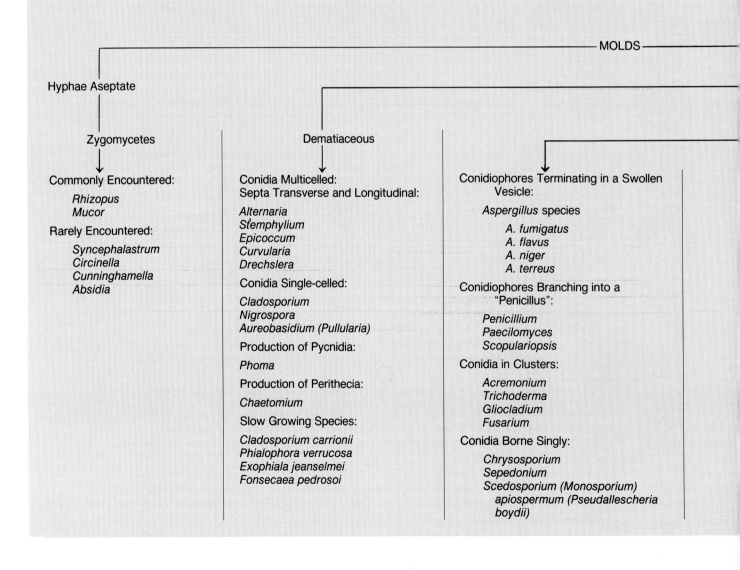

MOLDS

Hyphae Aseptate

Zygomycetes

Commonly Encountered:

*Rhizopus*
*Mucor*

Rarely Encountered:

*Syncephalastrum*
*Circinella*
*Cunninghamella*
*Absidia*

Dematiaceous

Conidia Multicelled:
Septa Transverse and Longitudinal:

*Alternaria*
*Stemphylium*
*Epicoccum*
*Curvularia*
*Drechslera*

Conidia Single-celled:

*Cladosporium*
*Nigrospora*
*Aureobasidium (Pullularia)*

Production of Pycnidia:

*Phoma*

Production of Perithecia:

*Chaetomium*

Slow Growing Species:

*Cladosporium carrionii*
*Phialophora verrucosa*
*Exophiala jeanselmei*
*Fonsecaea pedrosoi*

Conidiophores Terminating in a Swollen Vesicle:

*Aspergillus* species

*A. fumigatus*
*A. flavus*
*A. niger*
*A. terreus*

Conidiophores Branching into a "Penicillus":

*Penicillium*
*Paecilomyces*
*Scopulariopsis*

Conidia in Clusters:

*Acremonium*
*Trichoderma*
*Gliocladium*
*Fusarium*

Conidia Borne Singly:

*Chrysosporium*
*Sepedonium*
*Scedosporium (Monosporium) apiospermum (Pseudallescheria boydii)*

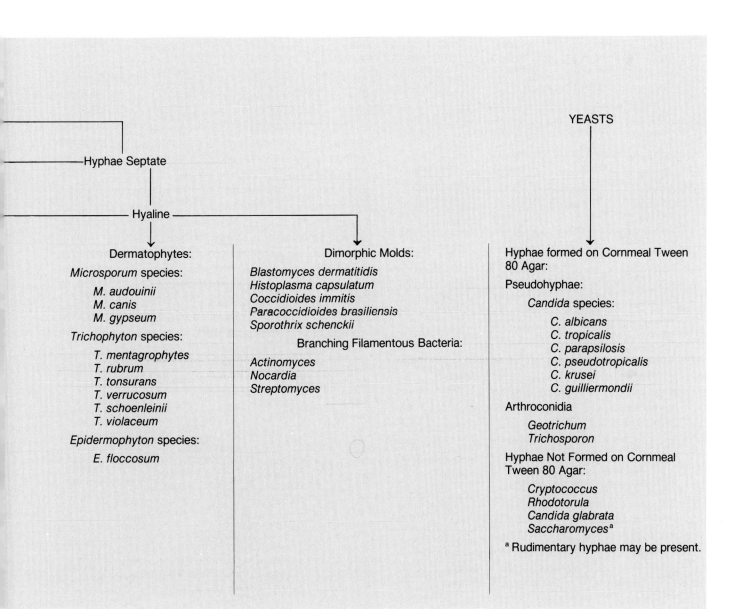

Hyphae Septate

Hyaline

YEASTS

**Dermatophytes:**

*Microsporum* species:

- *M. audouinii*
- *M. canis*
- *M. gypseum*

*Trichophyton* species:

- *T. mentagrophytes*
- *T. rubrum*
- *T. tonsurans*
- *T. verrucosum*
- *T. schoenleinii*
- *T. violaceum*

*Epidermophyton* species:

- *E. floccosum*

**Dimorphic Molds:**

- *Blastomyces dermatitidis*
- *Histoplasma capsulatum*
- *Coccidioides immitis*
- *Paracoccidioides brasiliensis*
- *Sporothrix schenckii*

**Branching Filamentous Bacteria:**

- *Actinomyces*
- *Nocardia*
- *Streptomyces*

**Hyphae formed on Cornmeal Tween 80 Agar:**

Pseudohyphae:

*Candida* species:

- *C. albicans*
- *C. tropicalis*
- *C. parapsilosis*
- *C. pseudotropicalis*
- *C. krusei*
- *C. guilliermondii*

Arthroconidia

- *Geotrichum*
- *Trichosporon*

**Hyphae Not Formed on Cornmeal Tween 80 Agar:**

- *Cryptococcus*
- *Rhodotorula*
- *Candida glabrata*
- *Saccharomyces*[a]

[a] Rudimentary hyphae may be present.

include a more in-depth study of basic mycology; however, it is also unfortunate that many individuals trained in classic mycology often have had no exposure to the clinical laboratory. And in reference to the medically oriented taxonomy, the medical technologist does not see the patient and is not in a position to assess the clinical manifestations of the disease; rather, encounters only the specimens that are submitted.

To assist medical technologists and microbiologists responsible for evaluating clinical specimens in their ability to identify accurately and to submit meaningful reports to the physician, we have suggested the use of a practical working schema which is designed to: (a) recognize the fungi recovered on artificial culture media that belong to a group known to be strictly pathogenic for man; (b) to recognize the fungi most commonly encountered in clinical specimens and (c) provide an easy to follow pathway by which a final identification can be made based on a few preliminary gross colonial and microscopic observations (see Table 5.3). It is assumed that the characteristics we selected to categorize the medically important fungi are valid; this assumption may be open to question but it is the best we have to go on based on our experience. It must be realized that microbiologists may take a certain degree of poetic license as far as pure taxonomy is concerned when identifying a given fungal species. The objective of the clinical microbiologist in naming a potential agent of disease is to supply one of many pieces of information required to make an accurate diagnosis so that appropriate treatment can be given to the patient. The practical working approach presented here provides a level of identification that is both pragmatic and functional; yet, it represents only one pathway among many. We should continue to strive to integrate all taxonomies in such a way that we derive a maximum understanding of the fungal agents and the diseases they cause.

## PRESUMPTIVE IDENTIFICATION OF PATHOGENIC FUNGI BASED ON PRELIMINARY OBSERVATIONS

The species of dimorphic fungi that regularly cause systemic mycoses in man often can be presumptively differentiated from the nonpathogenic or opportunistic (saprobic) fungi by observing several characteristics as listed below. It should be understood that these criteria are not absolute and may not be accurate in a small percentage of cases. Nevertheless, it is important that an unknown fungal isolate be recognized early since the treatment for mycotic diseases often is harsh, can be fraught with serious complications, and should not be given prematurely until a definitive diagnosis can be made. With experience, microbiologists will be able to recognize when these criteria do not apply with specific cultures and when additional observations must be made. Thus, the following criteria are only guides in making an initial assessment of the potential pathogenic nature of a fungal species recovered from clinical specimens:

1. The saprobic fungi are usually rapidly growing; they produce colonies within 1–5 days that are sufficiently mature to have produced the morphological structures necessary to make an identification. The dimorphic pathogenic fungi usually grow slowly; that is, mature colonies may not develop for 10 days or up to 45 days. However, strains of *C. immitis* commonly will produce mature colonies within 3–5 days and *H. capsulatum* and *B. dermatitidis* may also grow

in less than 10 days if there is a high concentration of organisms in the specimen. It must also be pointed out that *C. immitis* may grow on routine bacteriology culture media; therefore, microbiologists must exercise extreme care when examining any mold-like growth that may appear in culture plates and made examinations only under a properly functioning bacteriology safety hood.

2. The saprobic fungi cannot be converted to a yeast or spherule form by incubation at 37° C; by definition, the dimorphic fungi can. It should be mentioned that each of the dimorphic fungi has a saprobic counterpart that may appear microscopically similar and must be ruled out before a definitive identification is made. Thus, demonstrating the dimorphic nature of a clinical isolate or the presence of specific exoantigens is quite important.
3. Many species of the rapidly growing saprobic fungi produce brilliantly colored colonies due to the production of spores of various colors. The dimorphic fungi usually produce colonies that are white, buff, brown, gray, or gray-black. Some strains of *C. immitis* can produce pastel pink, orange, and yellow hues; however, other criteria will quickly be recognized to make the final identification.
4. Most saprobic fungi are inhibited in culture media containing cycloheximide; the dimorphic pathogenic species are able to grow in the presence of this antifungal agent. It is for this reason that the use of both inhibitory and noninhibitory isolation media should be used in setting up fungal cultures.
5. In general, the size of the hyphae of the dimorphic fungi are substantially narrower than those of saprobic fungi and tend to lie in parallel, giving a rope-like appearance. The delicate nature of the hyphae is what gives the colonies of dimorphic fungi a silk-like rather than woolly or cottony gross appearance.

## Practical Working Scheme for the Identification of Common Fungi Encountered in Clinical Laboratories

Using the criteria listed above, and applying the principles previously discussed on the gross and microscopic examination of fungal isolates, Table 5.3 is a practical working taxonomy for the identification of the fungi commonly encountered in the clinical laboratory. It is recognized that many species of fungi that may occasionally be encountered in the laboratory are not included in this list (68). This exclusion has been deliberate because we feel that students learning mycology for the first time and new clinical microbiologists should focus on knowing well those fungi that will comprise 99% of the species they will encounter in clinical laboratory practice. Identification of the less commonly encountered species often requires research beyond the scope of this primer and the reader is invited to consult the general texts listed in the Suggested Readings section. Once a less common species is identified, it can be included in one of the subgroups outlined in Table 5.3 if it is anticipated that this species will be encountered with some frequency in any given environment.

Table 5.3 also serves as an organizational outline by which the discussion of topics to be presented in the remaining portion of this text.

# PLATE 5.1

## MICROCULTURE TECHNIQUE

*A.* Use of sterile test tube to obtain agar blocks for use in the microculture technique.
*B.* Preparation of sterile humidity chamber using Petri dish containing 2% plain agar on top of which is placed a sterile microscope slide.
*C.* Alternate method for preparing microculture. Filter paper discs are placed in the bottom of the Petri dish and the microscope slide is supported by applicator sticks. The filter paper discs are kept moistened with sterile water during the incubation period.
*D.* Placement of agar block on surface of sterile slide.
*E.* Inoculation of the surface of an agar block in preparation of a microculture. Note that small portions of the fungal culture have been inoculated with a bent wire to four quadrants of the agar block.
*F.* Placement of sterile cover slip on the surface of the inoculated agar block prior to incubation.
*G.* Appearance of typical microculture after appropriate period of incubation. Note the mycelial growth under the surface of the cover slips.
*H.* Placement of the cover slip harvested from the microculture agar block on a drop of lactophenol aniline blue on a microscope slide in preparation for microscopic examination.

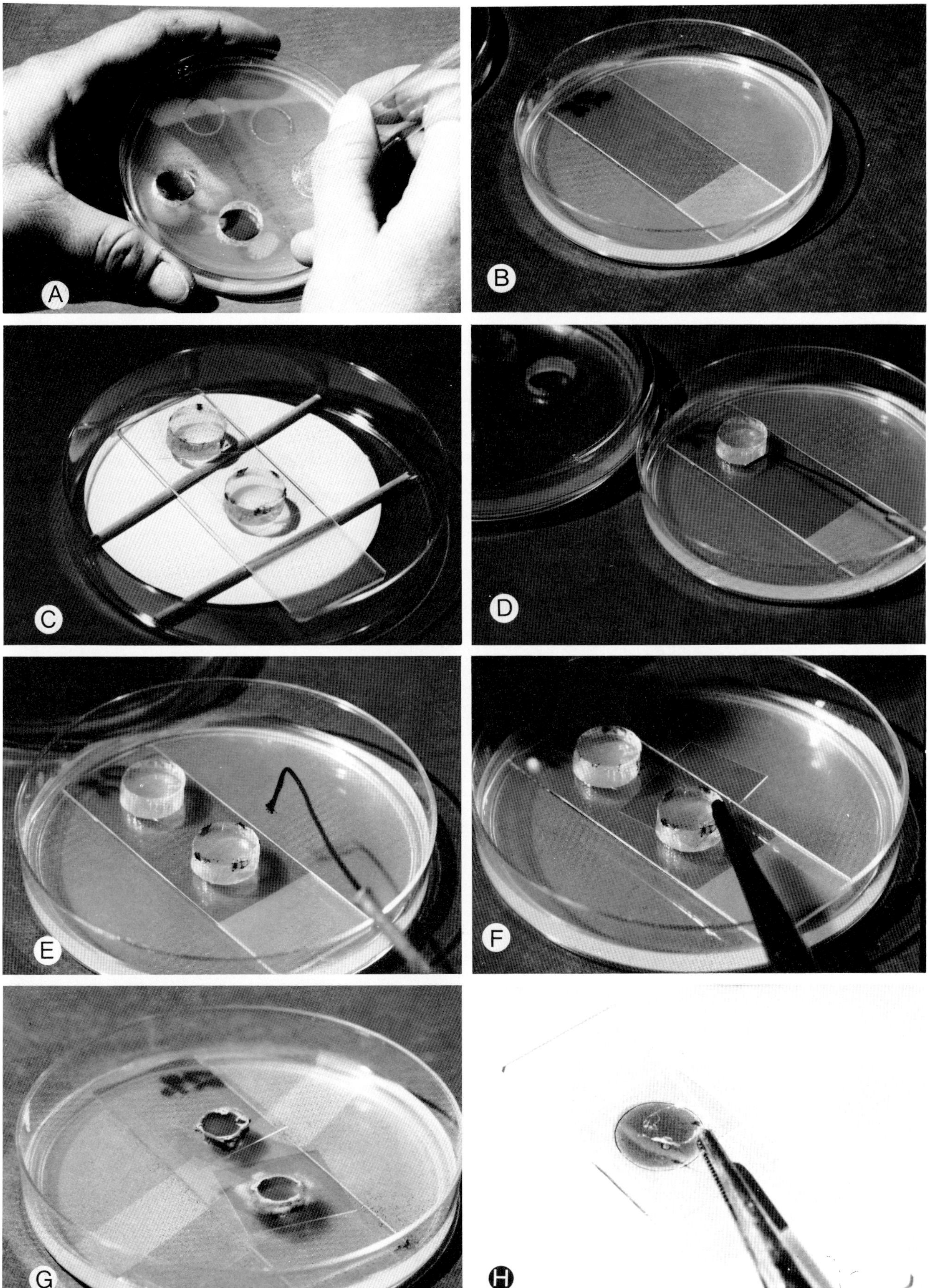
A
B
C
D
E
F
G
H

# PLATE 5.2

## GROSS APPEARANCE OF FUNGAL CULTURES

*A.* Surface of a fungal colony revealing a distinct entire margin and a fluffy or cottony consistency.

*B.* Fungal colony revealing a delicate, silky mycelium. A silky or delicate cobweb appearance is characteristic of the slower growing dimorphic fungi.

*C.* Glabrous, yeast-like rugose fungal colonies growing on the surface of blood containing culture medium, an appearance seen in particular with the slow-growing dimorphic fungi.

*D.* Typical appearance of smooth yeast colonies growing on fungal culture media. The colonies shown here also appear mucoid, a texture that suggests the capsular material of *C. neoformans*.

*E.* Close-in view of a fungal colony illustrating the prickly form of a dimorphic fungus. This type of colony is most commonly observed on blood containing medium during the process of conversion from the mold to the yeast form.

*F.* Fungal culture revealing colonies with a powdery or granular surface and radiating rugae. The granular texture and most intense pigmentation are seen in areas of heavy sporulation.

*G.* The growth shown here is described as woolly and the colony extends to the borders of the Petri dish without forming a margin. This colony type is typical of the Zygomycetes.

*H.* Typical appearance of a dematiaceous fungus illustrating the dark pigmentation characteristic of this group.

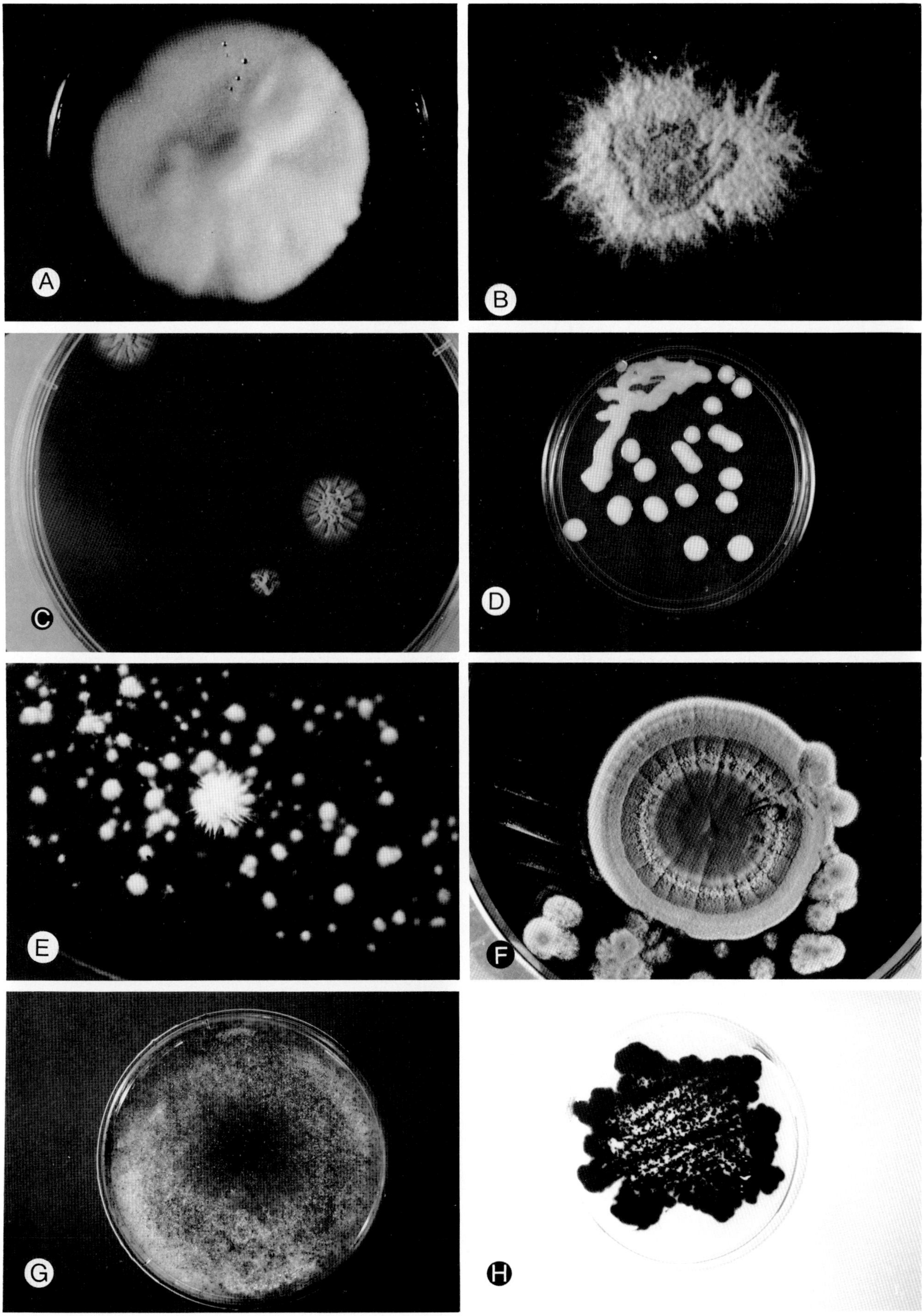
A
B
C
D
E
F
G
H

# PLATE 5.3

## MICROSCOPIC FEATURES OF FUNGAL CULTURES

*A.* Lactophenol-aniline blue preparation revealing broad, ribbon-like, aseptate hyphae characteristic of the Zygomycetes.
*B.* Septate hyphae of a hyaline mold.
*C.* Hyphae exhibiting racquet forms. These forms are commonly seen in cultures of dermatophytes and *C. immitis*.
*D.* Antler hyphae characteristically found in cultures of certain dermatophytes.
*E.* Pectinate body, a structure uncommonly seen in cultures of dermatophytes.
*F.* Spiral hyphae found in about 30% of cultures of *T. mentagrophytes* and less commonly in other dermatophytes.
*G.* Cleistothecium containing numerous ascospores representative of the sexual form of sporulation seen in the perfect stages of many fungi.
*H.* Group of sexually produced asci each containing clusters of 4–8 ascospores.

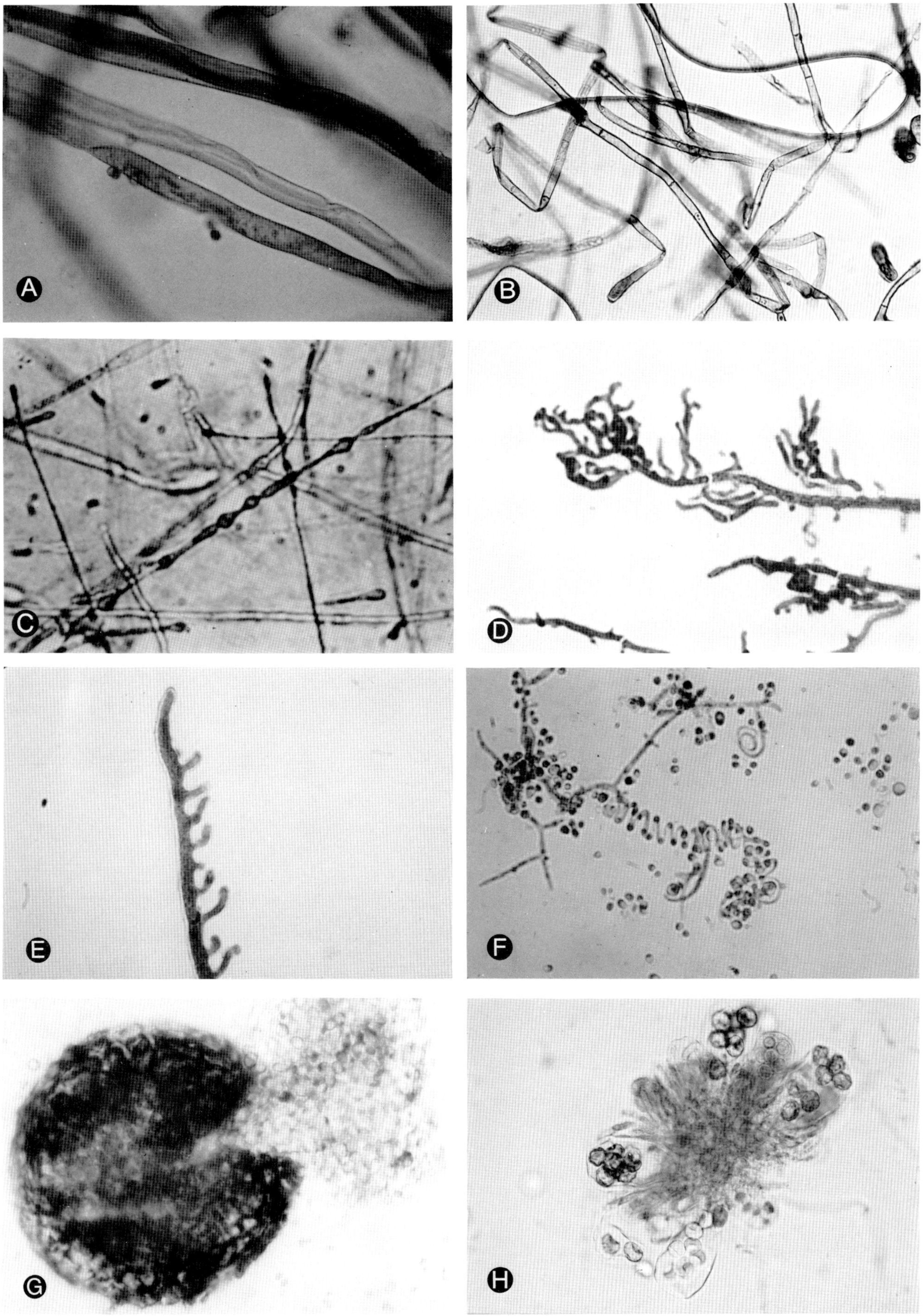
A
B
C
D
E
F
G
H

# PLATE 5.3 (*CONTINUED*)

## MICROSCOPIC FEATURES OF FUNGAL CULTURES

*I.* Pycnidium containing conidia. Compare the small size of the conidia seen at the periphery of this pycnidium to the larger size of the ascospores seen in Frame *G*.

*J.* Asexual pycnidia.

*K.* Appearance of arthroconidia as they typically appear in lactophenol-aniline blue mounts.

*L.* Appearance of budding yeast cells (blastoconidia) as they appear in direct mounts prepared from yeast cultures.

*M.* Pseudohyphae of a yeast colony resembling links of sausages in a chain. Pseudohyphae are differentiated from true hyphae which are usually considerably longer and have parallel walls with no indentations at the points of septation.

*N.* Intercalary chlamydospores most often observed in old fungal cultures.

*O.* Chains of chlamydospores as seen in some species of dermatophytes, particularly in mature cultures of *T. verrucosum* incubated at 35° C.

*P.* Terminal chlamydospore as it may appear in older cultures of certain saprobic or dermatophtic fungi.

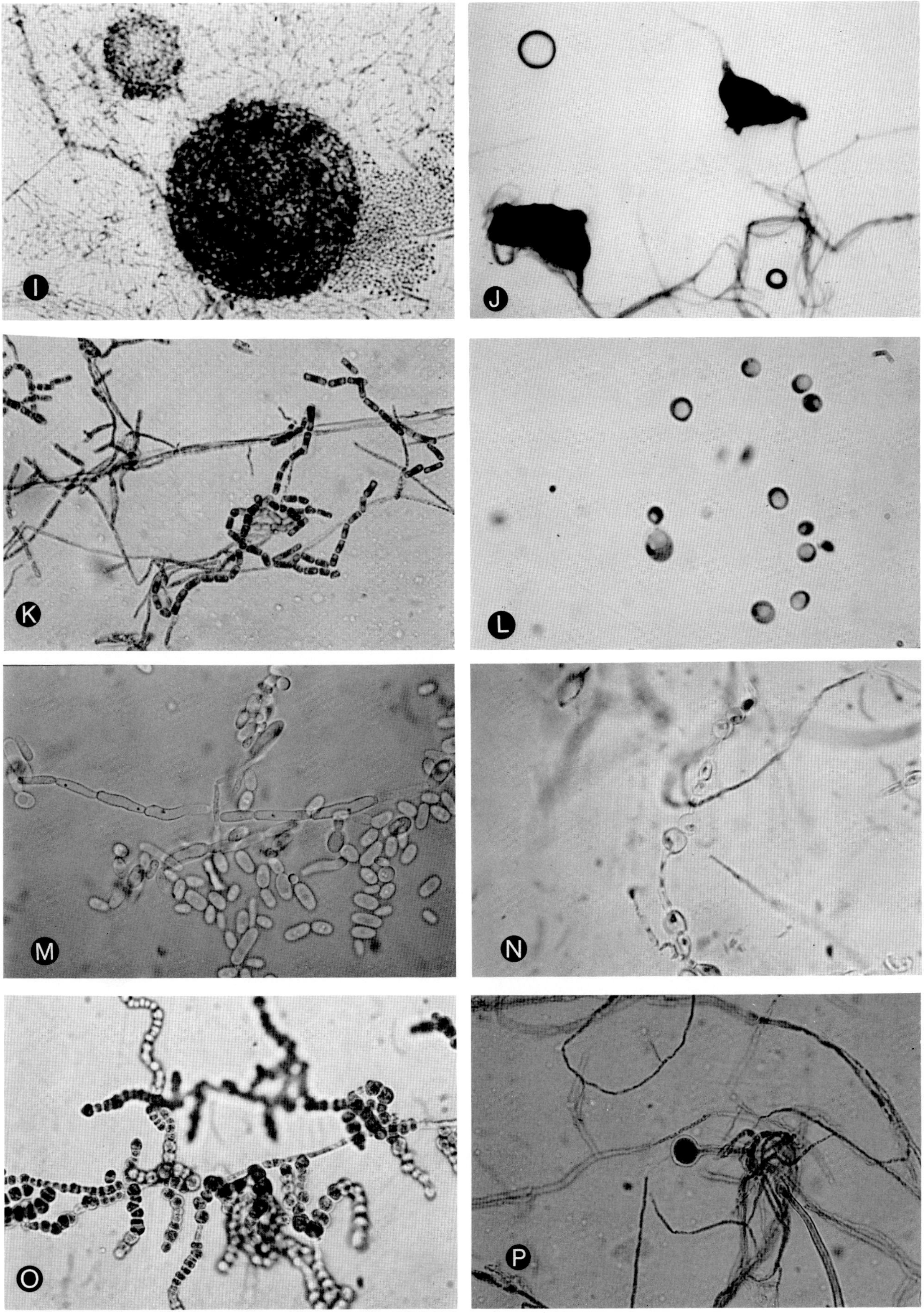
I
J
K
L
M
N
O
P

# PLATE 5.4

## STRUCTURES OF CONIDIAL SPORULATION

*A.* Photomicrograph of a lactophenol-aniline blue mount of a Zygomycete illustrating dark-staining sporangia supported by long sporangiospores. Low power.

*B.* Typical fruiting head of *Aspergillus* sp illustrating swollen vesicle covered by a row of phialides supporting the long chains of conidia. Low power.

*C.* Fruiting head of *Aspergillus* sp demonstrating a club-shaped vesicle, phialides radiating from the top portion of the vesicle and chains of conidia. High power.

*D.* Fruiting head of *Penicillium* sp illustrating a "penicillus," a structure composed of branching phialides simulating the fingers of a hand. Low power.

*E.* Appearance of the fruiting structure of *Acremonium* sp, one of several species of saprobes that produce conidia arranged in compact clusters. High power.

*F.* Photomicrograph illustrating the appearance of microconidia which in this culture are arranged in loose aggregates or dispersed singly along delicate hyphae strands. Note the presence of a single smooth walled, pencil-shaped macroconidium (*arrow*).

*G.* Appearance of multicelled, spindle-shaped, rough-walled macroconidia characteristic of *Microsporum* sp. A few widely scattered microconidia are seen in the background. High power.

*H.* Photomicrograph of a coremium characterized by the presence of several conidiophores cemented together to form a structure from which conidia are borne at the tips. High power.

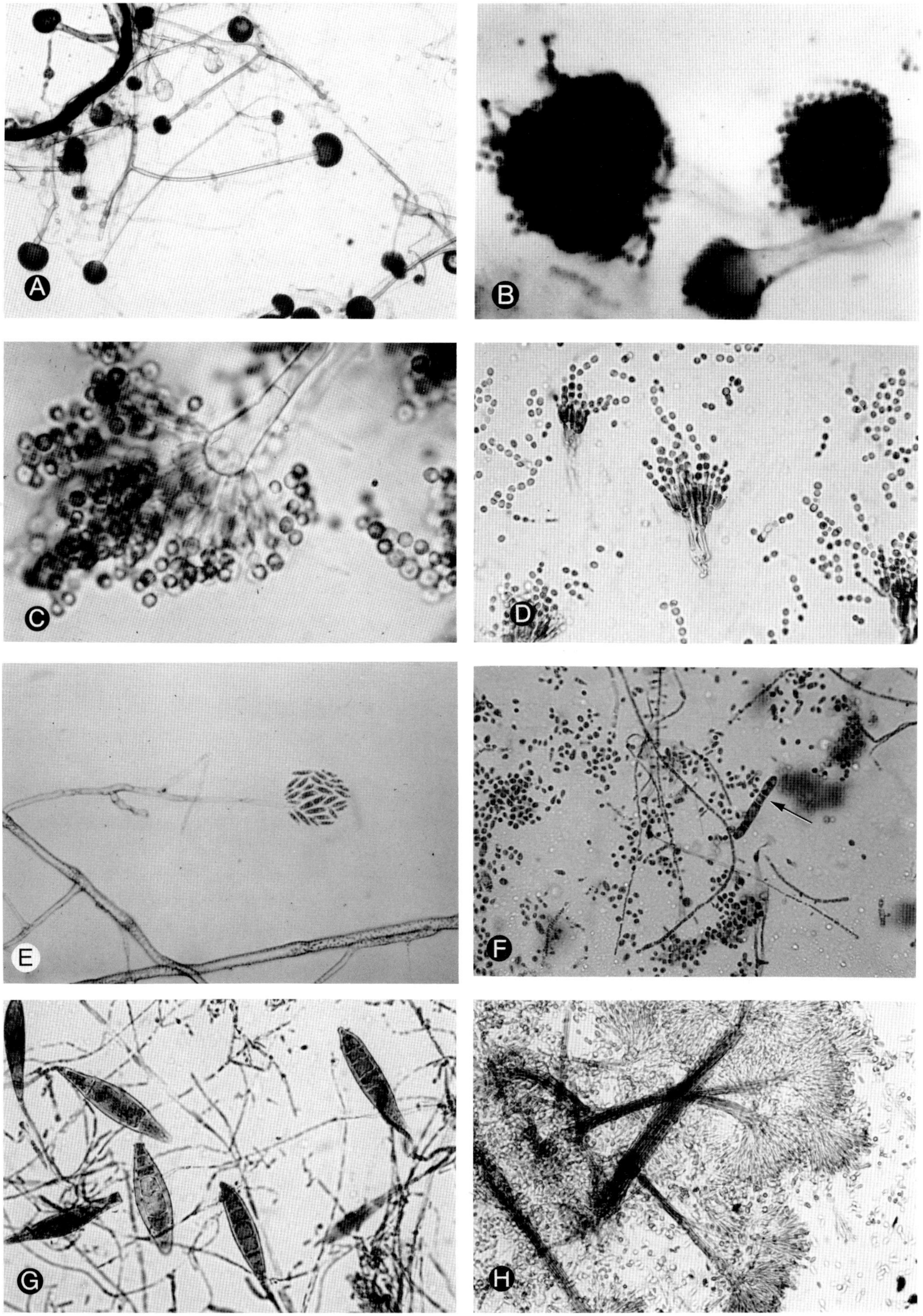
A
B
C
D
E
F
G
H

# Chapter 6

# LABORATORY IDENTIFICATION OF MOLDS

It is usually possible to determine if a fungal colony growing on culture media is a mold by direct visual examination. The filamentous nature of mold colonies usually is readily evident, although in some instances the hyphae may be so delicate that the colony may appear visually smooth or even yeast-like. In such cases, examination of a microscopic mount is necessary.

For many years the filamentous fungi have been considered either as nonpathogens or as contaminants. The term saprobes is currently being used for this group of fungi because of their ubiquitous presence in nature on decaying matter. Recently, it has been shown that saprobes previously thought to be nonpathogenic have produced severe, often fatal, infections in compromised patients. It is now necessary to consider all filamentous fungi as potential pathogens and care must be exercised when examining fungal cultures in the laboratory.

An evaluation of growth rate may be the most important observation when examining a mold. In general, the growth rate for strict pathogens, including *Histoplasma capsulatum, Blastomyces dermatitidis, Paracoccidioides brasiliensis,* and *Coccidioides immitis* is slow (1–4 weeks are required before colonies become visible), in contrast to the saprobes that form mature colonies within 1–5 days. In some instances *C. immitis* may produce visible growth within 2–5 days; however, most dimorphic pathogens recovered from culture specimens require an extended incubation period and it is recommended to hold all fungal cultures for 30 days before discarding.

As discussed in Chapter 5, the colonial morphology may be of little value in identifying the filamentous fungi because gross appearance can vary depending upon the culture medium used. For example, *H. capsulatum* on primary isolation often appears yeast-like on a blood-enriched medium, but on Sabouraud's dextrose agar or a similar medium, colonies appear white to tan and fluffy. Therefore, examination of microscopic mounts by the procedures outlined in Chapter 5 is required in most instances to make a genus or species identification. Recommended subculture media are listed in Table 4.3.

The purpose of this chapter is to provide a practical step by step approach to the identification of mold colonies that are recovered from clinical cultures in the laboratory. The sequence of presentation of the various groups of molds will follow the outline presented in the practical working taxonomy scheme shown in Table 5.3. An overview of Table 5.3 indicates that members of the Zygomycetes can be immediately split off from the other molds because they characteristically produce aseptate hyphae. The remaining molds that produce septate hyphae can be further subgrouped into those that are dematiaceous (dark) or hyaline (light). The hyaline group, in turn, includes three major subgroups, the opportunists, the dermatophytes and the dimorphic molds. The yeast-like organisms will be discussed in Chapter 7. The genus and species of fungi that are included within each of these major groupings can then be recognized by their unique microscopic characteristics, to be described in the text that follows and illustrated by black and white photomicrographs. The beginning student in mycology should find this approach valuable in the initial study of the molds, until instant recognition patterns become established through experience.

## *The Zygomycetes (Phycomycetes)*

A fungus recovered on laboratory culture media can be immediately suspected of belonging to the Zygomycetes by two easy to observe criteria: (a) the development of a fluffy, white, gray or brownish colony without a margin that diffusely covers the surface of agar within 1–2 days. The term "lid-lifters" has been used to describe this group of fungi because of their propensity for rapid and profuse growth (see Plate 5.2*G* and Figures 4 and 5 of Color Plate 4); and (b) the usually aseptate nature of the hyphae when observed microscopically (Plate 6.1*A*). Six genera of the Zygomycetes are recovered with varying frequencies in clinical laboratories:

Rhizopus
Mucor
Absidia
Circinella
Cunninghamella
Syncephalastrum

### Rhizopus and Mucor

All of these species have their source in the soil and have no special growth requirements. They all produce similarly appearing, very rapidly growing colonies that are fluffy or cottony and push up against the under surface of the lid of the Petri plate, initially white but fading into gray, brown, or black with maturity as spores are formed. ***Rhizopus*** and ***Mucor*** are the genera most frequently recovered from clinical specimens; the other species are only rarely encountered. The differentiation between *Rhizopus* and *Mucor* in particular and the classification of the other genera in general is more of academic interest than of clinical significance, but can be made easily by the observing the criteria described below.

The Zygomycetes are microscopically characterized by the features illustrated in Plate 6.1*A*. As a group, the hyphae are broad, ribbon-like, irregular in diameter, and devoid of septa (it should be pointed out that in more mature colonies, occasional septa may be seen both within the hyphae and more frequently in the sporangiophores). Identification is further confirmed by observing the presence of fruiting bodies in the form of sac-like structures called **sporangia** (singular = **sporangium**) within which are produced spherical, yellow or brown spores called **sporangiospores** (see Plate 6.1*B*). Each sporangium is supported by a special hyphal extension known as a **sporangiophore** (Greek *sporos* = "seed"; *angeion* = "vessel"; *phoros* = "bearing"). Each sporangiophore terminates in a globose swelling called the **columella** which is enclosed within the sporangium and from which the sporangiospores are derived. As the colony matures, the wall of the sporangium becomes fragile and finally breaks, releasing the myriads of sporangiospores into the environment to produce new colonies.

The members of some genera, notably *Rhizopus* and *Absidia*, produce root-like structures called **rhizoids** (Plate 6.1*C*). The genus *Mucor*, which does not produce rhizoids, can be excluded from consideration if rhizoids are identified in a culture. *Rhizopus* sp can be differentiated from the far less commonly encountered *Absidia* sp by noting the relationship between the rhizoids and the sporangiophores. Sporangiophores that arise from the stolon opposite the rhizoids are termed **"nodal"** in origin, a feature characteristic of *Rhizopus* sp (Plate 6.1*D*). If sporangiophores are derived **internodally** (Plate 6.1*E*), the genus *Absidia* is the most likely possibility. The genus *Mucor* should be considered when a culture that grossly is identified as a Zygomycete has sporangia, sporangiospores, straight sporangiophores, and the absence of rhizoids.

### Syncephalastrum

Microscopically, the fruiting heads of ***Syncephalastrum*** appear as daisy-like flowerettes in which elongated, cylindrical merosporangia are arranged radially around a spherical vesicle (Plate 6.1*F*). This appearance may sometimes be confused with the conidia-bearing fruiting head of certain species of *Aspergillus* (Plate 6.4). Differentiation can be easily made by noting that the spores of *Syncephalastrum* are indeed confined within merosporangia and that the hyphae are irregularly broad and aseptate in contrast to the fruiting head of *Aspergillus* sp which has chains of conidia eminating directly from the surface of vesicles and hyphae which are narrow, have parallel walls, and are septate. The rapidly growing, diffuse, gray-white to brown fluffy colony of *Syncephalastrum* also can be readily contrasted from the somewhat slower growing, powdery, and deeply pigmented colonies of most species of *Aspergillus* with their distinct outer margin of growth.

### Circinella

The microscopic features of ***Circinella,*** which is uncommonly encountered in the laboratory, are illustrated in Plate 6.1*G*. Globose sporangia usually packed with yellow-brown sporangiospores are borne from slender sporangiophores that characteristically loop back on themselves, a feature from which the genus name is derived (*circinus* = "circle").

### Cunninghamella

The fruiting head of the rarely encountered ***Cunninghamella* sp** is illustrated in Plate 6.1*H*. Spherical or oval spores, measuring 5–8 $\mu$m in diameter are borne individually from the surface of a globose to subglobose vesicle measuring between 30 and 60 $\mu$m in diameter. Each spore is individually attached to the surface of the columella by a pinched hair-like structure (denticle), which in composite gives the appearance of a "flowerette" (Plate 6.1*H*). In more mature colonies, the spores may develop echinulations giving a roughened appearance of the walls. Although true sporangia are not formed, *Cunninghamella* sp is nevertheless included with the Zygomycetes because the hyphae are broad and aseptate and because, under appropriate environmental conditions, some strains produce sexual zygospores.

## CLINICAL SIGNIFICANCE

The Zygomycetes produce human zygomycosis (also known as phycomycosis and mucormycosis), a disease that may assume several forms. The most serious and often fulminant form is **rhinocerebral** disease, with particular predilection for patients with uncontrolled diabetes or other diseases commonly associated with acidosis (3, 60). The primary infection is usually in one of the paranasal sinuses from which direct extension into the meninges may occur. Because the organism has a predilection for invading vascular channels, thrombosis and infarction of the infected tissue is common resulting in hemorrhage, infarction, and secondary spread to distant organs (60). Fungal ball infections may occur in the paranasal sinuses or in pulmonary cavities through the inhalation of air-borne spores.

Zygomycosis may manifest as **progressive pneumonia** primarily in patients with leukemia and lymphoma or as a primary gastrointestinal disease in malnourished individuals (3). Mucosal ulcerations and hemorrhagic infarction may occur in the gastrointestinal tract, leading to abdominal pain, hematemesis, and bloody diarrhea depending upon the site of involvement. **Primary cutaneous** zygomycosis may occur following direct inoculation of traumatized skin wounds or mucous membranes contaminated with soil that may lead to rapidly progressive cellulitis (12a). A recent outbreak of cutaneous zygomycosis has been reported in which several infections occurred after use of contaminated elastoplast tape (41).

*Rhizopus* sp and *Mucor* sp cause most human infections. Occasional infections caused by other Zygomycetes have been reported: a case of cutaneous (38) and fungus ball (43) infections with *Syncephalastrum* and cases of *Cunninghamella elegans* infections reported by Emmons and associates (21) and by Kwon-Chung and associates (48) in a patient with leukemia.

## THE DEMATIACEOUS MOLDS

A mold that from inception produces a colony with a dark brown, green-black, or black appearance of both the surface and the reverse sides, can be included in the dematiaceous group (Plate 5.2*H*). A variety of colonial forms are shown in Color Plate 1. Because the dark nature of the colony is easy to recognize visually, and can be quickly confirmed by observing dark yellow or brown septate hyphae on microscopic examination of a culture mount, beginners will have little difficulty in recognizing this group of fungi. Most species are rapidly growing saprobes, can be readily recovered from the air and soil and on rare occasions can cause invasive infections (phaeohyphomycosis) of the skin, lungs, brain, and other organs in immunosuppressed humans (22). These infections are to be differentiated from the subcutaneous lesions of chromoblastomycosis and mycetomas caused by the *Phialophora, Exophiala,* and *Wangiella* genera of black molds, a group that can be distinguished by their generally slower growth rate, their characteristic microscopic features to be described below and their more aggressive behavior (57, 58).

Many species of the dematiaceous molds are also incriminated in a variety of hypersensitivity conditions secondary to inhalation of air-borne spores. Bronchial asthma, farmer's lung, teapickers disease, maple bark strippers disease and bagassosis are among the conditions that atopic individuals may manifest upon contact with spores from various species of fungi, including members of the dematiaceous group (17, 29).

A practical subgrouping the dematiaceous fungi that will serve to aid in their identification is:

I. Rapidly growing dematiaceous molds:

A. Multicelled conidia with septations arranged transversely and longitudinally (dictyospores):

*Alternaria*
*Stemphylium*
*Epicoccum*
*Ulocladium*

B. Multicelled conidia with septations transversely only:

*Curvularia*
*Drechslera*
*Helminthosporium*

C. Single-celled conidia:

*Cladosporium*
*Nigrospora*
*Aureobasidium*

II. Slow-growing dematiaceous molds that produce conidia by phialides, annellides, in chains or sympodially:

*Fonsecaea*
*Phialophora*
*Wangiella*
*Exophiala*
*Cladosporium*

## MULTICELLED CONIDIA WITH HORIZONTAL/ LONGITUDINAL SEPTA

The first group of dematiaceous fungi to be discussed here are those characterized by the production of macroconidia that are divided by both longitudinal and cross-septa, resulting in a mosaic or muriform appearance. *Alternaria, Stemphylium, Epicoccum* and *Ulocladium* are the four genera to be discussed here.

### Alternaria

The colonies of ***Alternaria*** are dark gray-brown or green to black. A representative colony is shown in Figure 1 of Color Plate 1.

Microscopically the conidia are **muriform** (multicelled with both transverse and longitudinal septa) and have a dark brown color. They have an elongated drumstick-shape and may be arranged in chains, with the rounded, broad end of one conidia attached to the pointed, beak-like end of an adjacent one (Plate 6.2*A*).

*Alternaria* is found in the soil and is a plant pathogen. Occasional human cases of alternariosis have been reported: keratomycosis following trauma (6), cutaneous microabscesses in a farmer with leukemia (24), a case of osteomyelitis of the hard palate (including citations of 11 other cases of osteomyelitis due to *Alternaria*) (27), granulomatous pulmonary disease (54) and a case of invasive mycosis of the nasal septum caused by a mixed infection with *Alternaria* and *Curvularia* (55).

**Features of *Alternaria***

- Conidia have horizontal and longitudinal septa
- Conidia are borne in chains
- Conidia are golden brown

### Stemphylium

A representative colony is shown in Figure 3 of Color Plate 1. The colonies have a tendency to spread over the surface of the agar and the initial black colony may become overgrown with a smokey gray, sterile mycelium.

Microscopically, muriform, dark brown conidia resembling those of *Alternaria* are produced, except that the conidia of *Stemphylium* tend to be more oval or globose and are not arranged in chains but arise from the tips of simple or branched conidiophores (Plate 6.2*B*).

**Features of *Stemphylium***

- Conidia have horizontal and longitudinal septa
- Conidia tend to be spherical and are borne from the tips of single conidiophores

### Ulocladium

***Ulocladium*** is uncommonly recovered in the clinical laboratory; however, it is sometimes confused with *Stemphylium* since the conidia are similar in appearance.

*Ulocladium* is usually rapidly growing and produces colonies that are brown to olive to black and velvety. Some appear to be submerged beneath the surface of the medium. Conidia are produced sympodially; each develops behind and to one side of the other portion of the conidiophore. Conidia appear to be borne on a portion of the conidiophore that resembles a bent knee.

**Features of *Ulocladium***

- Conidia have horizontal and longitudinal septa
- Conidia are produced sympodially from "bent-knee" conidiophores

### Epicoccum

***Epicoccum*** may be suspected from the appearance of the colony, as illustrated in Figure 4 of Color Plate 1. The production of conidia with a variety of colors gives a variegated appearance to the surface of the colonies, with a play of yellow, orange, red and brown against a black background. The dark orange to brown pigment may diffuse into the agar.

Dark brown, globose, muriform conidia similar to those produced by *Stemphylium* are seen in microscopic mounts; however, the conidia of *Epicoccum* arrange in aggregates within dense masses of rebranching hyphal structures called **sporodochia** (Plate 6.2*C*). As the colony matures, the conidia may become roughened and wart-like on the surface.

**Features of *Epicoccum***

- Colonies have variegated appearance
- Conidia have horizontal and longitudinal septa
- Conidia commonly formed in aggregates

## MULTICELLED CONIDIA WITH TRANSVERSE SEPTA

The next group of dematiaceous fungi to be discussed are those that produce multicelled macroconidia that are divided only by transverse septa. *Curvularia* and *Drechslera* are the dematiaceous fungi from this group most commonly encountered in the laboratory. *Helminthosporium* species also belongs to this group but is only rarely encountered in clinical laboratories.

### Curvularia

The colony appears similar to that of *Alternaria* and is illustrated in Figure 1 of Color Plate 1. The genus name is derived from the configuration of the conidia, which have a boomerang appearance due to the rapid enlargement of one or more of the central cells (Plate 6.2*D*). The conidiophores are twisted and have a roughened blister or scar at the point where each conidium attaches in a sympodial arrangement.

The organism is primarily a soil saprobe, although occasional cases of human infections have been reported. Kaufman reported a case of endocarditis following cardiac surgery (40); separate cases of disseminated disease in football players originating in cutaneous sites of injury were reported by Harris and Downham (33) and by Rohwedder and associates (77); also Lampert et al. (50) reported a case of pulmonary and metastatic cerebral curvulariosis. In addition, ***Curvularia*** is sometimes a cause of mycotic keratitis as reported by Nityananda and associates (63).

**Features of *Curvularia***

- Conidia are golden brown
- Conidia are multicelled and curved with a central swollen cell

### Drechslera

The dematiaceous fungus belonging to the genus ***Drechslera*** was incorrectly identified as *Helminthosporium* sp in the previous editions of this text and by others. The colony is a blackish brown mold that is similar to *Alternaria* and *Curvularia.*

Microscopically, the smooth-walled, cylindrical conidia that are separated by transverse septa into four or more cells, each of which has the appearance of being vacuolated, closely resemble those of *Helminthosporium* species but are more rounded on the ends. The conidia of *Drechslera* sp, however, as described by Ellis (20) are sympodially produced first on one side, then the other from the successive tips of a continuously extending conidiophore, resulting in a twisted appearance (it may have been this wormlike appearance of the conidiophore that the designation *Helminthosporium* was originally derived—see Plate 6.2*E*). The conidiophores of *Helminthosporium* sp are not twisted and give rise to rows of conidia spaced laterally (Plate 6.2*F*).

*Drechslera* sp and *Helminthosporium* sp are both soil saprobes. A few human cases have been reported: a case of primary cutaneous infection with *Drechslera spicifera* (22), a fatal case of meningoencephalitis produced by *D. hawaiiensis* (25), a case of peritonitis in a patient receiving ambulatory peritoneal dialysis (*D. spicifera*) (66), a case of nasal obstruction and nasal bone invasion by *D. hawaiiensis* (88) and separate cases of *Drechslera* (91) and *Helminthosporium* keratomycosis complicating corneal ulcers (35, 45).

**Features of *Drechslera***

- Conidiophores twisted
- Conidia smooth-walled, cylindrical, septa transverse, 4 or more cells, produced sympodially

## SINGLE-CELLED CONIDIA

The following group of dematiaceous saprobes produce one-celled conidia, either in chains or singly. Of this group, *Nigrospora, Aureobasidium* and *Cladosporium* are the genera most commonly encountered in clinical laboratories.

### Nigrospora

Early colonies of ***Nigrospora*** appear hyaline; however, as the colony ages a distinct darkening of the hyphae can be observed and this fungus is usually included with the dematiaceous group. Initially the colony is white to light gray and spreads diffusely over the surface of the agar with a poorly defined margin. With time, the colony visibly darkens, due to the production of conidia.

The genus is easily identified by observing in microscopic mounts the jet black, globose, horizontally flattened, smooth, single-celled conidia that rest on top of short, broad, slightly twisted swollen conidiophores that tend to flare at the site of conidial attachment (Plate 6.2*G*).

**Features of *Nigrospora***

- Conidia black, globose, flattened horizontally, placed on top of a swollen conidiophore

### Aureobasidium (Pullularia)

***Aureobasidium*** can be suspected if a black yeast-like colony is observed (Figure 8 of Color Plate 1). Initially the yeast-like colony may be white to buff, creamy to mucoid but, in time, turns black with age. The distinctive vegetative hyphae and patterns of sporulation usually develop as the colony matures, making the differential identification possible. *Aureobasidium* has a distinctive microscopic morphology characterized by the formation of thick-walled, initially hyaline hyphae that turn dark brown as the pigmented arthroconidia develop (Plate 6.2*H*). Small, single-cell, oval conidia are produced directly from the hyphae singly or in clusters.

*Aureobasidium* is considered nonpathogenic (commonly mildew), although Vermeil and associates (86) have described one case of a cutaneous keloid-like lesion from which *Aureobasidium pullulans* was recovered.

**Features of *Aureobasidium***

- Thick-walled, dark brown arthroconidia
- Oval conidia produced singly or in clusters along hyaline hyphae

### Cladosporium

***Cladosporium*** is a rapidly growing saprobe frequently recovered from the air and is implicated in allergic respiratory disease. One species, *Cladosporium trichoides* (*bantianum*) has been incriminated as the cause of cerebral phaeohyphomycosis (also called cerebral chromomycosis), as originally described by Binford et al. (10) and more recently reviewed by Middleton and associates (59), who reported on two dozen cases culled from the literature. The condition is difficult to diagnose and most cases have been recognized only at autopsy. This organism is extremely hazardous and when suspected should be handled within a biologic safety hood. Kwon-Chung and associates have reported a case of pulmonary fungus ball infection produced by *Cladosporium cladosporioides* (47).

*Cladosporium carrionii* is one of the etiological agents of subcutaneous chromoblastomycosis. One may suspect this species if a slow-growing black mold is recovered from a subcutaneous lesion consistent with chromoblastomycosis with the microscopic morphology to be described below.

A variety of colonial types of the dematiaceous saprobes may be encountered, as illustrated in Figures 5, 6, and 7, of Color Plate 1. The colony types include variants that are deep green and velvety in consistency, gray brown or black and hairy, or dark gray to black, smooth and wooly. Microscopically, *Cladosporium* sp characteristically have dark brown, septate hyphae that branch freely. These hyphae support long chains of elongated, oval, one-celled conidia, with distinct dark terminal points of attachment, called disjunctors (Plate 6.3*A*). Cells with disjunctors are shield-shaped. These characteristics are helpful in identifying *Cladosporium* sp in microscopic preparations where fragments of spore chains have broken free from the main hyphal strand.

**Features of *Cladosporium***

- Colonies are velvety and darkly pigmented
- Conidia eliptical and in chains—some are shield-shaped due to disjunctors (scars)

## SLOW-GROWING DEMATIACEOUS MOLDS PRODUCING SINGLE-CELLED CONIDIA FROM PHIALIDES OR ANNELLIDES, ARRANGED IN CHAINS OR SYMPODIALLY

Although the dematiaceous molds most commonly recovered in the laboratory generally belong to the rapidly growing genera described previously, there are certain species, generally slower growing, that can be pathogenic for man. These pathogenic species are the agents of chromoblastomycosis and/or mycetoma of the skin and subcutaneous tissues, the incidence of which is quite low in the United States (28). Although the rate of growth is not the sole criterion by which one may recognize the dematiaceous pathogens, the appearance of a black mold within the time span of 10–30 days should alert the microbiologist that one of these species has been recovered.

The taxonomy of this group of fungi has not been totally resolved; however, the schema suggested by McGinnis and associates (57, 58) and also advocated by Rippon (70) will be briefly described here. The means for identifying this group of fungi have received only cursory treatment in most clinical laboratories because of their infrequent occurrence; yet, it is well for students to achieve some familiarity with their current names and morphological differences.

Formerly the fungi included in this group were divided into three genera based on differences in sporulation: (a) *Cladosporium,* which include those species that produce long chains of conidia via a form of sporulation called the cladosporium type (Plate 6.3*A*); (b) *Phialophora,* which include those species that produce short, flask-shaped **phialides**, usually with a well developed collarette from which clusters of conidia extrude from an apical pore (Plate 6.3*B* and *C*), a form of sporulation called the phialophora type; and, (c) *Fonsecaea,* a genus in which the types of sporulation are of a mixed type, but that uniquely includes those species in which conidia are borne singly at the ends and sides of denticles of swollen conidiophores (Plate 6.3*G* and *H*), known as the acrotheca type of sporulation. However, this genus separation proved to be too simplified and did not take into account other forms of conidiogenesis that has recently been revealed by scanning electron microscopic studies.

Under the classification proposed by McGinnis, ***Phialophora verrucosa*** and ***P. richardsiae*** are the only two species remaining in the genus *Phialophora.* Sporulation is purely of the phialophora type in which clusters of conidia exude from flask-shaped phialides that have a distinct collarette (Plate 6.3*C*). *P. verrucosa* produces phialides with collarettes that are flask-shaped, differing from *P. richardsiae* in which the collarettes are more saucer-like. *P. jeanselmei* is the former name of a dematiaceous fungus known to be associated with human mycetomas. The taxonomic placement of this species has been a problem because the type of sporulation differs from other *Phialophora* sp in that the conidiophores are long and slender and terminate in a tapered tip where a series of growth rings called **annellides** are observed which, in turn, give rise to a succession of conidia (Plate 6.3*D*). The annellides are best seen using an oil immersion objective; however, they are often difficult to distinguish even by experienced mycologists. This species is now placed in the genus *Exophiala* and is called *E. jeanselmei*, a designation that also subsumes that former species *P. gougerotii.*

In addition to the microscopic characteristics just described, **E. jeanselmei** often appears as a black yeast early in its growth. Microscopically, darkly pigmented, single-celled budding yeasts are the only cellular forms seen at this stage of development (Plate 6.3*E*). Similar yeast forms may also be produced by members of the genus ***Wangiella.***

The dematiaceous fungus now called ***Wangiella dermatitidis*** (formerly *Fonsecaea dermatitidis*) has microscopic features similar those of *E. jeanselmei.* Early in colonial development, the colony may be yeast-like and microscopically only dark-staining, single-celled yeast forms may be seen. As the colony matures, a low velvety aerial mycelium forms, usually having a deep olive-green to green-black pigmentation. Microscopically, long tube-like phialides, tapered at the ends and giving rise to clusters of single-celled conidia at their tips are seen (Plate 6.3*F*); however, these differ from those of *E. jeanselmei* by being devoid of annelides.

The genus ***Fonsecaea*** now includes two species, ***F. pedrosoi*** and ***F. compacta,*** both associated with human chromoblastomycosis. The latter species is only rarely encountered in clinical laboratories. Sporulation is usually a mixture of combinations of the cladosporium, phialophora, and acrotheca types, although one tends to predominate depending upon the strain isolated. The mixture of acrotheca and cladosporium types as shown in Plate 6.3*G* and the sympodial arrangement of conidia seen in Plate 6.3*H* are prevalent in most cultures. *F. compacta* is characterized by the formation of conidia in compact heads; however, this species is rarely encountered in the United States.

The genus ***Cladosporium,*** in which sporulation is purely of the cladosporium type, includes two species, ***C. carrionii*** and ***C. trichoides.*** These species are also associated with chromoblastomycosis and phaeohyphomycosis of the brain. *C. trichoides* can be distinguished from *C. carrionii* by the formation of conidia that are longer (5–7 $\mu$m vs. 4–5 $\mu$m) and more elliptical. *C. trichoides* does not hydrolyze casein which *C. carrionii* does, a helpful differential feature when the morphological features are too subtle to identify a given isolate definitively.

For a full description of the morphology of these fungi and for detailed illustrations of the differences in conidiogenesis that form the basis for the genus and species designations, see the texts of McGinnis and Rippon as cited in the Suggested Readings section. A summary of the species designations and the identifying characteristics of this group of dematiaceous fungi is as follows:

**Differential Characteristics of Pathogenic Dematiaceous Molds**

| | |
|---|---|
| *Phialophora verrucosa:* | Tube-like or flask-shaped phialides each with a flask-shaped collarette |
| *Phialophora richardsiae:* | Flask-shaped phialides with a saucer-like collarette at the tip |
| *Exophiala jeanselmei:* | Dark-staining, single-celled yeast forms present in young cultures; long tapering conidiophores with annellides (rings of growth) sometimes visible under oil immersion |
| *Wangiella dermatitidis:* | Yeast forms seen in early cultures; long tube-like or flask-shaped phialides without a collarette |
| *Fonsecaea pedrosoi:* | Cladosporium type of sporulation; phialides with collarettes may be present; conidial heads with sympodial arrangement of conidia, with primary conidia giving rise to secondary conidia |
| *Fonsecaea compacta:* | Morphology similar to *F. pedrosoi* except that conidia are borne in compact heads |

| | |
|---|---|
| *Cladosporium carrionii:* | Cladosporium type sporulation with long chains of elliptical conidia (2–3 μm × 4–5 μm) borne from erect, tall, branching conidiophores |
| *Cladosporium trichoides:* | Morphologically similar to *C. carrionii* except conidia are somewhat longer (5–7 μm) and more elliptical; differentiation must be made biochemically: *C. trichoides* does not hydrolyze casein, *C. carrionii* does |

## THE HYALINE MOLDS

As shown in the practical working taxonomy in Figure 5.3, the hyaline group of molds, which have septate hyphae that are transluscent or colorless, includes (a) species of rapidly growing saprobes that commonly cause opportunistic infections in humans; (b) the dermatophytes which infect the keratinized portions of the integument including skin, hairs, and nails, and (c) the dimorphic pathogens that are inherently virulent and cause serious and often fatal deep seated mycoses in man.

The rapidly growing hyaline saprobes can be suspected in culture by observing colonies, usually with a well-defined border, in which the surface may be deeply pigmented and is granular in consistency from the production of large numbers of conidia. The dermatophytes are somewhat slower growing, tend to show less surface pigmentation, but may produce water soluble yellow, orange, or red pigments that diffuse into the agar, producing a brightly colored reverse surface. The dimorphic pathogens most commonly are slow growing (10–30 days), produce silky colonies that are initially white but may turn dull gray, brown, or buff with age. They grow on enriched media containing chloramphenicol and cycloheximide and can be converted to a yeast or spherule form by incubation at 37 °C.

### Identification of the Rapidly Growing Hyaline Molds

Including four species of *Aspergillus*, 14 species of hyaline molds will be discussed here, which comprise the majority of this group of fungi that will be encountered in the clinical laboratory. To assist the beginning student in developing a working orientation, the hyaline saprobes can be divided into the following subgroups based on differences in the microscopic appearance of their fruiting bodies:

| | |
|---|---|
| Conidiophores terminate in swollen vesicles; conidia borne in chains (catenulate), acrogenously from the tips of phialides covering portions or all of the surface of the vesicle (Plate 6.4) | *Aspergillus* |
| Conidiophores freely branching; conidia borne in chains from the tips of annellides or phialides borne on metulae forming a "penicillus" (Plate 6.5) | *Penicillium*<br>*Paecilomyces*<br>*Scopulariopsis* |

| | |
|---|---|
| Conidiophores single; conidia borne in clumps or clusters, generally at the tips of straight or flared conidiophores (Plate 6.5) | *Acremonium*<br>*Trichoderma*<br>*Gliocladium*<br>*Fusarium* |
| Conidiophores single, each bearing a solitary conidium (Plate 6.5) | *Chrysosporium*<br>*Sepedonium*<br>*Scedosporium (Monosporium) apiospermum (Pseudallescheria boydii)* |

## CONIDIOPHORES TERMINATING IN SWOLLEN VESICLES

### *Aspergillus* species

**Aspergillus** sp are the rapidly growing molds encountered most frequently in clinical laboratories and should be considered in the differential identification of any rapidly growing hyaline mold. Due to the profuse production of pigmented spores, many species of *Aspergillus* display colonies with vivid colors, ranging from blue to green to yellow. Pure white and jet black colonies may also be encountered.

Microscopically the genus is characterized by chains of small or oval to spherical conidia borne in chains from the tips of phialides radially positioned over the surface of the swollen tip of the conidiophore, called the vesicle (Plate 6.4). Some species of *Aspergillus*, notably *A. glaucus* and *A. nidulans*, may reproduce sexually in culture media and **cleistothecia** may be observed (Plate 6.4*G*). Cleistothecia in turn contain saclike **asci** which contain **ascospores,** the morphology of which can also be used to make species identification for those who have the experience. Also rarely seen are Hülle cells which are hyaline spherical structures of unknown origin (Plate 6.4*H*).

Several hundred species of Aspergilli have been described. Based on differences in colony pigmentation, the size and length of the conidiophores, the shape of the vesicles, the presence of metulae, the position of the phialides, the size and length of chains of the spores and other criteria, Raper and Fennell (69) have classified the Aspergilli into 18 groups. The ability to recognize and correctly classify the various species of *Aspergillus* is an art that requires considerable experience and is both beyond the scope of this text and beyond clinical significance as well. Fortunately the clinical microbiologist need only become thoroughly familiar with the characteristics of three species that are clinically significant, ***A. fumigatus, A. niger,*** and ***A. flavus.*** A fourth species, *A. terreus*, is also being recovered with increasing frequency from clinical specimens (49, 81) and a brief description will be included here.

*Aspergillus* sp are widespread in nature where they act as common saprophytes on grains, leaves, in soil, and in manure. Conidia readily become dispersed and man most commonly becomes infected by inhaling airborne spores (4). Thus, various forms of pulmonary disease, including fungus ball infection, allergic bronchopulmonary and invasive pneumonitis are the more common manifestations; metastatic disseminated aspergillosis involving various organs, otomycosis, mycotic keratitis, onychomycosis, endocarditis, and mycetomatous skin diseases are less common clinical variants (80).

### *Aspergillus fumigatus*

This mold grows in 2–6 days on media free of cycloheximide. The colony grows with a distinct margin, is fluffy and white during early growth but later becomes velvety or sugary and blue-green to gray-green from the production of pigmented conidia. The reverse of the colony is white to cream in color.

Microscopically, the hyphae are hyaline and septate. The conidiophores are moderate in length and have a characteristic foot cell at their bases. The vesicles are dome-shaped and the upper one-half to two-thirds of the surface is covered with a row of phialides from which long chains of 2–3 μm in diameter, globose, echinulate conidia are borne, with a tendency to sweep inward toward the center (Plate 6.4*C* and *D*). Cultures can withstand tempratures up to 45° C.

***A. fumigatus*** is the species most commonly causing opportunistic infections in humans including granulmatous pulmonary, allergic bronchopulmonary and disseminated disease forms, the latter in immunocompromised patients (4).

**Features of *Aspergillus fumigatus***

- Green or green-gray, sugary colonies
- Uniseriate—phialides over upper one-half of a hemispherical vesicle
- Chains of globose conidia tending to sweep inwardly

### Aspergillus niger

This fungus begins as a white colony that may turn yellow, but soon develops a black pepper effect on the surface as conidia are produced, which may become so dense in time to produce a black matte (Fig. 3 of Color Plate 4). The reverse of the colony remains buff or cream-colored, a feature differentiating ***A. niger*** from the dematiaceous molds.

Microscopically, the hyphae are septate and the conidiophores are long and smooth. The vesicle is spherical and gives rise to large metulae and smaller phialides from which dense clusters of jet black conidia are produced, obscuring the surface of the vesicle (Plate 6.4*A* and *B*).

*A. niger* is a common cause of cavitary fungus ball lesions of the lungs or nasal sinuses. It also is incriminated in a high percentage of cases of otomycosis, where positive cultures can be obtained from the draining, crusted lesions of "swimmers ear" (83).

**Features of *Aspergillus niger***

- Colonies black
- Vesicles spherical
- Biseriate—large metulae and smaller phialides
- Conidia black and roughened

### *Aspergillus flavus*

Colonies grow rapidly within 5 days at 25° C and have a well developed, cottony aerial mycelium that is some shade of yellow, brown or yellow-green when mature (Fig. 2 of Color Plate 4).

Microscopically, the vesicles measure from 20–60 μm, are globose and sporulation occurs over the entire surface. The phialides arise from the vesicle (uniseriate) or from a primary row of metulae that give rise to phialides on the vesicle (biseriate). The phialides in turn give rise to short chains or globular masses of 3–5 μm in diameter yellow or yellow-orange elliptical to spherical conidia that become echinulate with age (Plate 6.4*E*).

***A. flavus*** is a common cause of allergic pulmonary aspergillosis and can be recovered from sputum or from bronchial mucous plugs (19, 70). The presence of numerous eosinophils and Charcot-Leyden crystals in microscopic mounts of sputum is highly suggestive of *A. flavus* infection. *A. flavus* may also cause disseminated disease in immunocompromised hosts.

**Features of *Aspergillus flavus***

- Colonies usually yellow or yellow-green
- Vesicles round, sporulation over entire surface
- Phialides alone (uniseriate) or with metulae (biseriate) may be present

### Aspergillus terreus

The colony grows rapidly at 25° C and within 5 days becomes cinnamon-buff or brown and cottony initially but becoming sugary as profuse sporulation develops. Radial or irregular rugae may form.

Microscopically, the vesicles are small, averaging about 15 μm, domelike and support metulae and a row of phialides and short chains of elliptical conidia measuring 2–3 μm in diameter (Plate 6.4*F*). In addition, single aleuriospores are produced on submerged hyphae.

***A. terreus*** is causing clinical disease in man with increasing frequency. Seligsohn and associates (81) reported a case of osteomyelitis and cited four additional cases in which *A. terreus* caused infection of a nasal polyp, cervical lymph node, meninges and disseminated disease. Chronic illness and debilitation, break down of local barriers and corticosteroid therapy are etiological factors. Laham and Carpenter (49) recently reported *A. terreus* as the causative agent of cases of endocarditis and bronchopulmonary disease. *A. terreus* is a major cause of otomycosis in Japan and Formosa (21).

**Features of *Aspergillus terreus***

- Colonies cinnamon-colored
- Vesicles dome-like
- Metulae and phialides both present (biseriate)
- Aleuriospores produced on submerged hyphae

## HYALINE MOLDS FORMING A PENICILLUS

Three genera that characteristically form a brush-like penicillus (Latin—Penicillus = "brush") fruiting body are *Penicillium, Paecilomyces*, and *Scopulariopsis.*

### Penicillium

A representative colony is shown in Figure 2 of Color Plate 2. Usually shades of green, blue-green or green-brown are observed; however, as illustrated in Figure 3 of Color Plate 2, yellow and brown colonies may also be seen. The surface of the colony is velvety to powdery from the dense production of conidia and radial rugae are often formed. Often drops of exudate may be present on the surface of colonies.

Microscopically, the characteristic feature is the brush-like branching of the conidiophores resembling the fingers on a hand (Plate 6.5*A* and *B*). Long chains of small, spherical conidia are borne from flask-shaped phialides on top of the branched metuale. The blunt, sawed-off appearance of the terminal portions of the phialides is a helpful characteristic in differentiating *Penicillium* from *Paecilomyces*, the latter forming long tapered phialides (Plate 6.5*C*).

***Penicillium*** sp only rarely cause human infections, although a wide spectrum of infections have been reported: a deep-seated infection from *Penicillium marneffei* (16); a case of pleural effusion (26), a case of post surgical endocarditis following placement of a prosthetic valve (32), a solitary pulmonary penicillium granuloma (53), and additional sporadic cases reported in the texts by Emmons et al. (21) and Rippon (70).

**Features of *Penicillium* species**

- Colonies usually green or green-yellow
- Conidial heads resemble brushes or fingers
- Phialides with blunt ends giving rise to chains of conidia

### Paecilomyces

The colonies have a distinct margin and initially are smooth but become velvety to powdery as sporulation develops. Colonies may be white or have light pastel shades of pink, violet, yellow-brown, or gray-green. Some strains produce colonies that are similar to *Penicillium* on visual examination. A yellow-brown variant is shown in Figure 4 of Color Plate 2.

Microscopically, ***Paecilomyces*** is characterized by a penicillus similar to that seen with *Penicillium*; except, that the phialides are long and tapering, ending in a sharp point (Plate 6.5*C*). In some instances, phialides are borne singly along the hyphae. Small spherical to elliptical conidia are borne from the tips of the tapered phialides, in chains that are usually shorter than those seen with *Penicillium.*

Several human infections caused by *Paecilomyces* species have been reported in the recent literature. Prosthetic valvular endocarditis was reported by Haldane and associates (30), infection in a patient receiving a cadaveric kidney transplant by Harris and associates (34), cases of endophthalmitis (one following implantation of an intraocular lens and the other following traumatic penetration of the eye with a nail) have been reported by Mosier et al. (61) and by Rodriques and MacLoed (76), respectively, a case of chronic maxillary sinusitis was reported by Rockhill and Klein (75) and a cutaneous infection of the cheek by Takayasu and associates (84). Thus, no particular syndrome is caused by this fungus; rather, infections occur in instances where conidia gain entrance into a wound or site of injury. Cases in which mycotic diseases are presumably transferred with organ transplants is a potentially serious situation because transplant recipients receive immunosuppressive therapy to prevent host vs. graft rejection.

**Features of *Paecilomyces* species**

- Conidial heads resemble a small delicate penicillus
- Phialides are tapered toward their tips.

### Scopulariopsis

A typical colony of ***Scopulariopsis*** is shown in Figure 5 of Color Plate 2. Characteristically the colonies are some shade of yellow, buff or brown and have a powdery surface from the profuse production of conidia, interrupted by irregular radial rugae.

Microscopically, annellides form a simple or branching penicillus (Greek—*scopula* = "broom") that bear chains of lemon-shaped conidia that are 2–2½ times the size of those produced by *Penicillium* (Plate 6.5*D*). As the colony ages, the conidia have a propensity to become roughened or echinulated with a flattened base (Plate 6.2*E*). *Scopulariopsis* has caused septicemia and hypersensitivity pneumonitis in drug addicts. *Scopulariopsis* conidia are widely distributed in plant juices and stems and have been found to contaminate crude opium preparations. A pulmonary hypersensitivity disease similar to maple bark strippers disease has been reported (29). A case of deep-seated scopulariopsosis has been reported by Sekhon and associates (82) and cases of onychomycosis have been cited by Rippon (70). This fungus is also commonly recovered from nail scrapings.

**Features of *Scopulariopsis***

- Penicillus composed of single or branching annellophores
- Relatively large conidia, echinulate with a flattened base

## HYALINE MOLDS FORMING CONIDIA IN CLUSTERS

Some molds produce a mucilaginous substance during conidiogenesis that causes the conidia to be held in clumps or clusters. The more commonly encountered genera having this characteristic are: ***Acremonium*** (previously ***Cephalosporium*), *Gliocladium*, *Trichoderma*, and *Fusarium*.** The conidiophores in each of these species are commonly quite delicate in structure and the clusters of conidia are easily detached when making microscopic mounts. However, the tendency for the conidia to adhere into small clusters can still be observed, a helpful feature in making a preliminary identification.

### Acremonium (Cephalosporium)

The young colonies are moist and almost yeast-like in appearance because the aerial mycelium is so delicate (Fig. 6 of Color Plate 2). White, pink, and gray-yellow colonial variants are seen.

Microscopically, the conidiophores (phialides) of ***Acremonium*** are long, delicate, and almost hair-like in appearance. One-celled, elliptical conidia, arranged in irregular clusters simulating the surface of the cerebral cortex of the brain (an appearance in keeping with the former name, *Cephalosporium*, meaning "head spore") are supported at the tips of the conidiophores (Plate 6.5*F* and *G*).

Rippon (70) cites several cases in which *Acremonium* species have been recovered from specimens of patients with infections: cerebrospinal fluid (18), renal grafts (65), skin, nails, blood, pleural fluid and gastric contents. The organism is also one of the known causes of mycotic keratitis.

**Features of *Acremonium***

- Long narrow phialides with clusters of single-celled oval to cylindrical conidia at their tips.

### Trichoderma

A characteristic colony is shown in Figure 7 of Color Plate 2. A border does not form; rather, the growth is diffuse over the surface simulating a lawn. A yellow, yellow-green, or green cottony mycelium, tending to become more powdery with age, is commonly observed. The development of a yellow lawn should suggest *Trichoderma*; the lawn-like colony of *Gliocladium* sp tends to be green.

Microscopically, the short, flask-shaped, tapered phialides simulate those seen with *Paecilomyces*, except a penicillus is not formed and conidiophores occur singly at obtuse angles to the hyphae. Small, elliptical to globose conidia are in clusters (Plate 6.5*H*).

***Trichoderma*** sp are found in soil and are considered to be nonpathogenic. Cases of septicemia following infusion of contaminated intravenous dextrose have been cited by Beneke and Rogers (7).

**Features of *Trichoderma***

- Short conidiophores with tapered, flask-shaped phialides
- Oval to spherical conidia clustered at tips of phialides

### Gliocladium

This hyaline saprobe is infrequently encountered in clinical laboratories, but can be quickly suspected when observing a diffuse green, fluffy to granular lawn-like colony on the agar surface.

Microscopically, this fungus is relatively easy to identify because of the compact clusters of spherical conidia that are supported on the tops of phialides arranged in a penicillus (Plate 6.5*I* and *J*). This fungus is encountered as a laboratory contaminant and is not pathogenic for humans.

**Features of *Gliocladium***

- Clusters of conidia situated on top of a well-developed penicillus

### Fusarium

This fungus can also be suspected when recovered in culture media because of the propensity to produce a delicate lavender, purple, or rose-red pigments that color both the mycelium and the reverse side of the agar (Fig. 8 of Color Plate 2.8). Yellow variants may also occasionally be seen. The colony consistency is usually cottony or woolly with less tendency to become granular.

Microscopically, ***Fusarium*** is one of the few saprobic fungi that produce both macroconidia and microconidia. Single-celled microconidia are borne in small clusters from the tips of short phialides, similar to the configuration described for *Acremonium* species (Plate 6.5*K*). The identification of *Fusarium* can be confirmed by observing the characteristic large boat-shaped or sickleform, multicelled macroconidia (Plate 6.5*L*).

*Fusarium* species are the most common cause of mycotic keratitis (23, 90). Endophthalmitis may result in cases where there is penetration of the orbit, such as the case caused by a thorn as reported by Rowsey and associates (78). *Fusarium* species have also been recovered with some frequency from skin lesions in burn patients (1, 87), or in cases of traumatic cutaneous infections (13). Osteomyelitis of the knee secondary to trauma (11), onychomycosis (79), and disseminated disease in a patient with malignant lymphoma are other isolated cases that have been reported (89).

**Features of *Fusarium***

- Presence of macroconidia and also microconidia in some isolates
- Macroconidia are cylindrical, multicelled, and sickle-shaped
- Microconidia are arranged in clusters on top of short, delicate phialides

Two other hyaline saprobes, although uncommonly encountered in clinical laboratories, are important to keep in mind because they have microscopic features similar to the mold forms of *B. dermatitidis* and *H. capsulatum* and certain strains may also microscopially suggest one of the dermatophytes. If the source of the culture is not known; or, if the specimen is listed as "skin scraping" or "lesion of skin," the chance for a misidentification may even be greater. Therefore, when a rapidly-growing hyaline mold producing single conidia, colloquially known as "lollypops" is seen, additional information about the clinical presentation of the patient and the exact source of the culture will be helpful in making the final identification.

*Chrysosporium* sp and *Sepedonium* sp are the saprobic counterparts to *B. dermatitidis* and *H. capsulatum*, respectively; however, they should be relatively easy to differentiate by their more rapid growth rate, the inability to convert them into a yeast form when incubated at 35° C or inability to produce specific exoantigens and by their inhibition by media containing cycloheximide.

### Chrysosporium

The colony matures rapidly on antibiotic free fungal media, usually within 4–6 days. The colony appearance is not distinctive, having initially a white, fluffy appearance that may turn buff or tan with age.

Microscopically, single-celled, globose to clavate conidia are borne singly from short, delicate conidiophores produced laterally from the hyphae (Plate 6.5*M*). Numerous arthroconidia are also produced. It is of interest to know that *B. dermatitidis* was once known as *C. dermatitidis*.

As mentioned above, the saprobic species of ***Chrysosporium*** can be differentiated from *B. dermatitidis* by their more rapid growth rate, their inability to grow on fungal culture medium containing cycloheximide and their incapability of converting to a yeast form when incubated at 35° C on enriched media or to produce specific exoantigens. *Chrysosporium* sp is not pathogenic for man.

#### Features of *Chrysosporium*

- Single-celled conidia produced on short to long simple conidiophores
- Colony rapidly growing; no yeast form at 37° C or specific exoantigen produced

### Sepedonium

The colonial morphology of ***Sepedonium*** sp is similar to that of *Chrysosporium*, forming nondistinct white or gray, cottony colonies.

Microscopically, microconidia may be rarely produced which are single-celled, fusiform, and borne singly from delicate conidiophores that are difficult to demonstrate in microscopic mounts.

More characteristic of the genus is the production of 10–15 μm in diameter, spherical tuberculate macroconidia, closely simulating the macroconidia produced by the mold form of *H. capsulatum* (Plate 6.5*N*). The echinulations on the surface of the macroconidia of *Sepedonium* tend to be more delicate, shorter and not as sharply spiked as those produced by *H. capsulatum*; however, the differences are subtle and not sufficiently clear to make a differential identification on this basis alone. The inability to convert to a yeast form or produce precipitin reactions in the exoantigen test should be used to confirm the identification. *Sepedonium* sp are not pathogenic for man.

#### Features of *Sepedonium*

- Single-celled, rough-walled round to oval conidia on single conidiophores
- Colony rapidly growing. No yeast form at 37° C or specific exoantigen produced

### Pseudallescheria

***Pseudallescheria (Petriellidium) boydii*** is a hyaline mold that is the most common cause of eumycotic mycetoma endemic in the United States (28). Colonies are typically woolly and have a distinctive mousey gray color with a dark brown or brown-black reverse side. *Pseudallescheria boydii* is the perfect stage of *Scedosporium (Monosporium) apiospermum*, the latter term derived from the appearance of the elliptical sperm-shaped, single-celled conidia borne singly from the tips of short or long conidiophores (Plate 6.5*O*).

The perfect stage, *P. boydii*, may also reproduce asexually in a similar fashion, but may also form sexually derived cleistothecia (Plate 6.5*P*). If cleistothecia are observed in an unknown culture, the differentiation from *B. dermatitidis* can be easily made; however, in their absence, the formation of "lollypops" by both species may be microscopically confusing. If the cleistothecia are observed, the organism should be named *P. boydii*; however, when only conidia are observed, it should be called *Scedosporium apiospermum* according to current recommendation of many taxonomists.

The natural habitat for *P. boydii* is in the soil and the organism has a worldwide distribution. Man contracts the subcutaneous mycetomatous form of the disease when traumatic wounds or breaks in the skin become contaminated with soil containing infective spores. Individuals who have repeated contact with sewage sludge, polluted streams, and manure of poultry and cattle are also at increased risk. Pseudallecherial mycetomas have been resistant to antibiotic therapy and surgical excision or amputation in more severe cases has been necessary for successful treatment.

The majority of infections are confined to the subcutaneous tissue; however, recently a wide variety of infections involving other organs have been caused by *P. boydii*, most commonly in patients with chronic debilitating disease or who are immunosuppressed. Pulmonary lesions include bronchial and pulmonary colonization, fungus ball formation in pre-existent lung cavities and invasive pseudallescherial pneumonia. The pathology and clinical manifestations can closely simulate pulmonary aspergillosis. The organism has also been recovered from brain abscesses.

Infections of the nasal sinuses and septum can simulate rhinocerebral zygomycosis; or, fungus ball involvement of a paranasal sinus cavity may occur. Cases of meningitis, arthritis, endocarditis, mycotic keratitis and otomycosis secondary to *P. boydii* have also been reported. For a more complete review of these various syndromes and tabular citations of cases culled from the literature, refer to Rippon (70).

**Features of *Pseudallescheria***

- Colonies have a mousey gray appearance
- Single-celled, brownish conidia produced singly or in small groups at the tips of simple single conidiophores (*Scedosporium*)
- Some cultures produce cleistothecia containing ascospores (*Pseudallescheria*)

## THE DERMATOPHYTES

The dermatophytic fungi are most unique in that they require and utilize keratin for growth; therefore, infections with this group of fungi are confined to the superficial integument including the outer stratum corneum of the skin, the nails, and the hairs. Cutaneous infections caused by this group of fungi have historically been called "ringworm" because the classical lesions of tinea corporis tend to assume a circular form. The word "tinea," a term used almost since antiquity to describe small insect larvae which at one time was thought to be the cause of this form of skin disease, is still retained today when describing the clinical manifestations of conditions that are more properly designated as dermatophytosis.

Since the advent of griseofulvin and topical antifungal compounds, interest in the recovery and identification of the dermatophytic fungi has waned in clinical practice. Therefore, although infections with dermatophytes represent the most common type of fungal disease in humans, many clinical laboratories receive cultures infrequently. Nevertheless, medical and epidemiological considerations dictate that clinical microbiologists and laboratory technologists remain familiar with the culture characteristics of the dermatophytes so that exact identifications can be made when necessary.

Perhaps more important than making an exact species identification of a dermatophyte recovered from a clinical specimen is to prevent mistaking the cutaneous extension of a dimorphic fungus with a tinea infection, or visa versa. Careful examination of the clinical lesions when possible and direct microscopic examination of the specimens or of stained slides is usually sufficient to make the correct diagnosis.

### Clinical Types of Tinea Infections

Several well-defined clinical syndromes are produced by various species of dermatophytes, as outlined in Table 6.1. Each disease type has one or more common etiological agent and the lesions vary to some degree depending upon the part of the body affected and the species of dermatophyte involved. Dermatophytes have a geographic distribution; for example, *M. ferrugineum* is found predominantly in Japan, *T. concentricum* in the South Pacific, *T. schoenleinii* in northern Europe, and *T. violaceum* in the countries surrounding the Mediterranean. These fungi are rare in the United States. In contrast, *T. mentagrophytes, Epidermophyton floccosum*, and *T. rubrum* are virtually universal in distribution and cause tinea pedis (athlete's foot) among other dermatophytoses throughout the world. *T. tonsurans,* formerly an uncommon isolate in the United States, is currently being recovered with increasing frequency as a cause of tinea capitis infections in immigrants from Mexico and Latin America. Table 6.2 lists the frequency with which several species of dermatophytes have been encountered in the mycology laboratory of the Mayo Clinic during the period 1977–1984.

Dermatophytoses endemic in animals that can be transmitted to man are called zoophilic (*M. canis* from dogs and cats, *T. verrucosum* from cattle, for example); those acquired from the soil are termed geophilic (*M. gypseum*) while those derived by direct contact or indirectly through fomites from other humans are termed anthropophilic (*E. floccosum, M. audouinii, T. rubrum, T. tonsurans, T. schoenleinii*, and *T. violaceum*). Infected combs and brushes, towels, bed linens, undergarments, locker room floors and theater chair backs are modes of transfer for many of the anthropophilic dermatophytes. This difference in modes of transfer is one reason why species identification may still be important.

Following is a simple outline of the genera of dermatophytes and the parts of the human integument typically affected:

**Infection Sites for Dermatophyte Genera**

| | |
|---|---|
| *Epidermophyton floccosum:* | Skin, rarely nails, hair never |
| *Microsporum* sp: | Hair and skin, rarely nails |
| *Trichophyton* sp: | Hair, skin and nails |

**Table 6.1.** Etiological Agents and Clinical Manifestations of Dermatomycoses

| Type | Dermatophyte | Type of Infection | Diagnostic Procedure |
|---|---|---|---|
| *Tinea capitis* (Ringworm of the Scalp) | *Microsporum audouinii* *Microsporum canis* | "Gray patch ringworm," ectothrix, alopecia, intense itching; *M. audouinii* contagious; *M. canis* produces more inflammation | Ultraviolet light (Wood's lamp): apple-green fluorescence |
| | *Trichophyton tonsurans* *Trichophyton violaceum* *Trichophyton schoenleinii* | Endothrix—"black dot ringworm" due to hairs broken off at or beneath the scalp surface; tendency to chronic infection into adult life with development of exophytic, crusted lesions known as favus | Endothrix infected hairs do not fluoresce KOH mounts and/or culture |
| | *Trichophyton mentagrophytes* *Trichophyton verrucosum* | Ectothrix—inflammatory reaction prominent; suppurative folliculitis leading to chronic exophytic lesions known as kerions | Wood's lamp reaction is negative KOH mounts and/or culture |
| *Tinea corporis* (Ringworm of the Body) | *Trichophyton rubrum* *Trichophyton mentagrophytes* | Dry, scaly, spreading in annular or ring forms, healing centrally Moist, vesicular (iris) type with suppurative folliculitis | Skin scrapings from leading edge adjacent to normal skin |
| *Tinea cruris* (Ringworm of the Groin) | *Epidermophyton floccosum* *Trichophyton rubrum* *Trichophyton mentagrophytes* | Spreading or serpiginous, swollen area of inflammation usually with an elevated, eczematous margin, bilaterally but not symmetrically on inner thighs; *T. mentagrophytes* infections more inflammatory | Pick vesicle from eczematous lesion and identify fungus in KOH mount; culture from serpiginous margin |
| *Tinea barbae* (Ringworm of the Beard) | *Trichophyton mentagrophytes* *Trichophyton verrucosum* | Scaling, spreading lesion with a vesiculopustular border; deep pustular lesions with kerion formation seen in chronic cases | Demonstrate hyphae in KOH mounts of exfoliated skin scales or culture |
| *Tinea pedis* (*Tinea manuum*) (Ringworm of feet and hands) | *Trichophyton mentagrophytes* *Trichophyton rubrum* *Epidermophyton floccosum* | Chronic dermatitis between toes; peeling and fissuring of skin Hyperkeratotic, dry type with silver white scales on soles or palms; vesiculation and pustule formation may be complications | Demonstrate hyphae in KOH mounts of exfoliated skin scales or fluid from pustules; or, culture positive |
| *Tinea unguium* (Ringworm of the nails) | *Trichophyton rubrum* *Trichophyton mentagrophytes* | Small, yellow spot begins at base of nail spreading over nail which in chronic infections becomes brittle, friable, and distorted | Demonstrate hyphae in KOH mount of nail shavings or nail bed detritus |

**Table 6.2.** Dermatophytes Recovered at Mayo Clinic, 1977–1984

| Organism | Number |
|---|---|
| *Epidermophyton floccosum* | 329 |
| *Microsporum audouinii* | 3 |
| *Microsporum canis* | 25 |
| *Microsporum gypseum* | 11 |
| *Trichophyton mentagrophytes* | 803 |
| *Trichophyton rubrum* | 3031 |
| *Trichophyton tonsurans* | 13 |
| *Trichophyton verrucosum* | 120 |
| *Trichophyton violaceum* | 1 |

### *Tinea Versicolor* (pityriasis versicolor)

A superficial cutaneous infection that should be mentioned in relationship to dermatophytosis is tinea versicolor, the name given to a relatively common, superficial fungal infection of the skin caused by ***Malassezia furfur***, a lipophilic fungus described by Malassez in 1874. This organism is difficult to recover in vitro, a practice not attempted in most laboratories because a culture medium including sterile oil or a fatty acid is required. The diagnosis is easily made by demonstrating the two characteristic fungal forms in KOH mounts of exfoliated skin scales: (a) short, 3 μm in diameter, occasionally branched abortive hyphae and (b) clusters of spherical, budding yeast cells measuring between 4 and 8 μm in diameter (Plate 6.6*A* and *B*). This microscopic picture has colloqually been called "spaghetti and meatballs." This can be distinguished from *Candida* sp that exhibit a mosaic pattern when yeast cells are arranged in close proximity.

The species name *furfur* is based on a Latin stem meaning "bran," a general term used in classical medicine to refer to the dandruff-like scaling of the skin. The term versicolor refers to the clinical appearance of the skin lesions that present as macular, scaly spots ranging in color from yellow-brown to dark-brown depending on the complexion of the patient. The infection most commonly involves the superficial skin of the anterior chest or trunk. Thus inflammation or pruritis are absent. Infections most commonly occur under conditions of poor hygiene, are self-limited and easily cured by repeated and thorough washing of the affected areas with soap.

### Recovery and Early Recognition of Dermatophytes from Clinical Specimens

Most species of dermatophytes grow well on nonselective medium such as Sabouraud's dextrose agar; or will grow on commercial media such as Mycosel (BBL) and Mycobiotic (Difco) that contain chloramphenicol and cycloheximide. Cultures should be incubated at room temperature and should not be refrigerated as *E. floccosum* is susceptible to chilling. Dermatophyte test medium (Pfizer), formulated to select dermatophytes from other molds, yeast, or bacteria that may be recovered from clinical specimens, are used primarily in physician office laboratories. The medium is designed to select for the growth of dermatophytes and make detection easy by the appearance of a red color from conversion of the phenol red indicator in the presence of growth. Most saprophytic fungi, yeasts, and bacteria either do not grow on the medium or produce no color change. However, since several saprobic fungi may produce a color change in this medium, microscopic examination of any growth should always be made. Clinical microbiologists should become familiar with this medium since cultures are often referred for evaluation.

In general, the macroscopic appearance of dermatophyte colonies is not useful in making a species identification. Color Plate 3 includes photographs of a number of representative dermatophyte colonies, each described on the face sheet opposite the color photographs. Colonies may appear fluffy or granular depending upon the density of sporulation. *Microsporum* sp. and *E. floccosum* often produce yellow, buff, or cinnamon-colored colonies; *T. mentagrophytes* or *T. rubrum* can be suspected if a water-soluble, wine-colored pigment is observed on the reverse surface of the colonies. Because the dermatophytes have a particular propensity to become sterile when grown on an artificial culture medium, transfer from fluffy tufts of growth should be avoided when making microscopic mounts for study or when subculturing to other media.

The examination of KOH preparations of clinical specimens was described in Chapter 3 and the detection of hyphae in skin scrapings or spores of infected hairs should correlate with a positive culture. In their absence, several vegetative forms such as racquet hyphae, spiral hyphae, favic chandeliers (antler hyphae), and pectinate bodies may offer the first clue that the culture is a dermatophyte. The genus and species identification, however, depends upon the detection and morphological recognition of conidia and their arrangement on the hyphae. There are few biochemical characteristics available for identifying the dermatophytes but most of the species seen in the clinical laboratory do not require their use.

### Laboratory Differentiation of the Dermatophytes

Table 6.3 illustrates and lists the criteria by which the imperfect forms of the dermatophytes are first divided into one of three genera, ***Epidermophyton, Trichophyton,*** and ***Microsporum.***

The genus ***Epidermophyton*** is characterized by the presence of large clavate, multisegmented smooth-walled macroconidia, commonly borne either singly or in clusters of two or three from the tips of short conidiophores. Chlamydospores may be numerous in older cultures. Microconidia are not produced, an important differential feature in separating *E. floccosum* from *Trichophyton* sp and those strains of *Microsporum* in which the roughened walls of the macroconidia are not prominent (*M. nanum*, for example).

The genus ***Trichophyton*** is characterized by the presence of many microconidia borne either laterally along the hyphae **(en thyrses),** or in clusters **(en grappe).** Microconidia are generally quite numerous, except for some of the slower growing species, such as *T. violaceum, T. schoenleinii*, and *T. verrucosum*, cultures of which are virtually always devoid of sporulation. Macroconidia are only sparsely produced by most strains of Trichophyton recovered in the laboratory; when present, they are cigar-shaped, borne from short, delicate conidiophores and have a thin, smooth wall without evidence of roughening or echinulations.

The genus ***Microsporum*** is characterized by the production of large, multiseptate, rough-walled macroconidia that are derived singly from short conidiophores or attached by means of a "break-away" cell that distintegrates when the spore is released. The macroconidia tend to be more spindle-shaped than the elongated, cylindrical macroconidia of *Trichophyton* sp. Microconidia are usually sparsely produced, except for occasional strains where they may be quite numerous. *M. audouinii* is an exception in that macroconidia and microconidia are rarely produced; rather, favic chandeliers, chlamydospores, and other vegetative forms are all that suggest that an unknown culture might be a dermatophyte.

Identification of the dermatophyte species most commonly recovered in the clinical laboratory is shown in Table 6.4. This approach is practical and should be

**Table 6.3.** Genus Differentiation of the Dermatophytes

| *EPIDERMOPHYTON* | *TRICHOPHYTON* | *MICROSPORUM* |
|---|---|---|
| 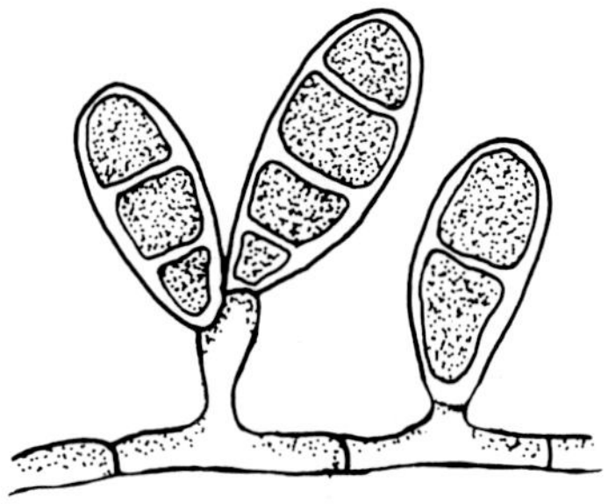 | 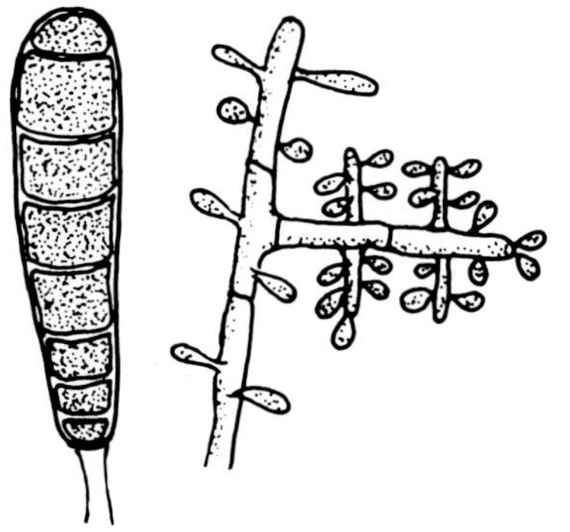 | 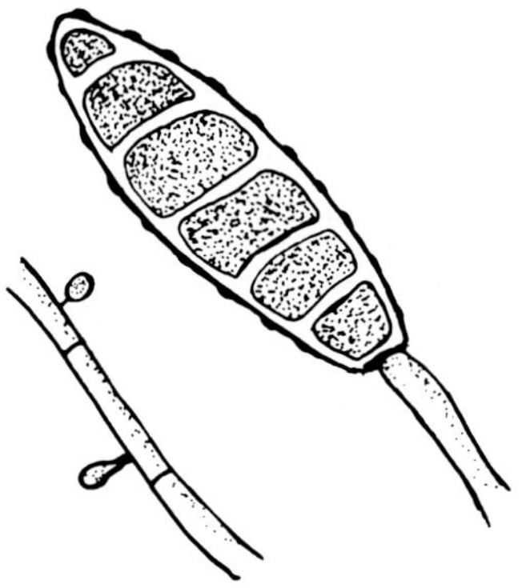 |
| Macroconidia are divided into two or four cells, are smooth-walled, and borne singly or in clusters of two or three from short conidiophores.<br>Microconidia are not produced. | Macroconidia are thin-walled with a smooth surface, pencil- or fusiform-shaped, and divided into three to eight cells. They are not usually produced by *T. violaceum* or *T. schoenleinii.*<br>Microconidia are generally numerous and borne singly along the hyphae (*en thyrses*) or in grape-like clusters (*en grappe*). | Macroconidia are divided into three to seven cells, are barrel- to spindle-shaped and have thick, roughened, or spiny walls.<br>Microconidia are few in number, globose to tear-shaped and borne in single, sessile fashion from the hyphae. |

useful to students learning the rudiments of mycology. Of the some three dozen currently recognized species of dermatophytes, less than one dozen are clinically significant worldwide and only the seven species reading from left to right in Table 6.4 are important in the United States.

Following is a brief orientation to the use of Table 6.4. The dermatophytes can be divided into three genera based upon the microscopic appearance of the macroconidia and microconidia:

### Dermatophyte Genus Identification: Epidermophyton

Macroconidia are large, smooth-walled, club-shaped and borne singly or in clusters of two or three
Microconidia are not produced

### Epidermophyton floccosum

The genus ***Epidermophyton*** is characterized by the presence of large, clavate, multisegmented, smooth-

**Table 6.4.** Dermatophyte Identification Schema*

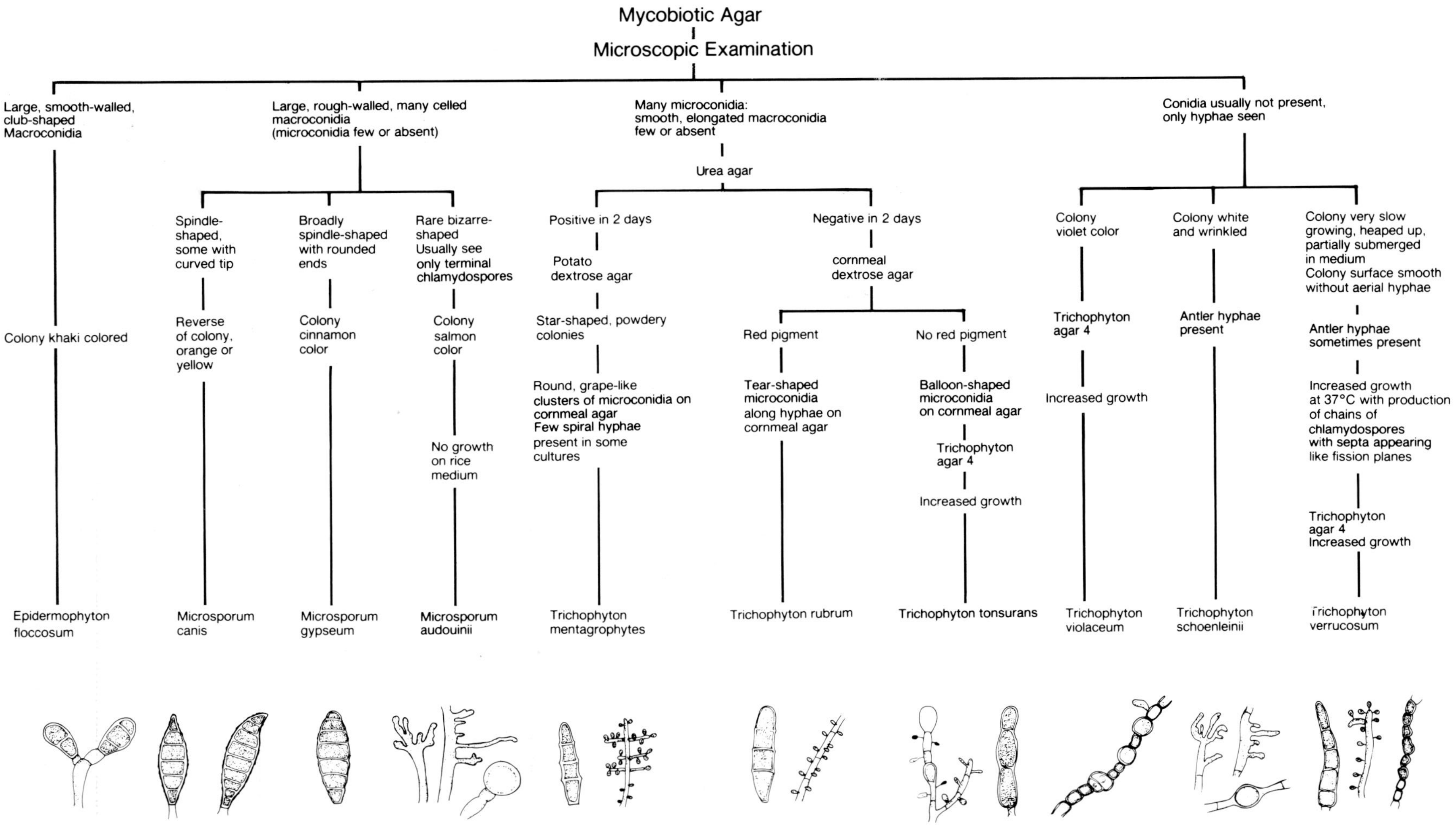
Mycobiotic Agar
Microscopic Examination
Large, smooth-walled, club-shaped Macroconidia
Colony khaki colored
Epidermophyton floccosum
Large, rough-walled, many celled macroconidia (microconidia few or absent)
Spindle-shaped, some with curved tip
Reverse of colony, orange or yellow
Microsporum canis
Broadly spindle-shaped with rounded ends
Colony cinnamon color
Microsporum gypseum
Rare bizarre-shaped Usually see only terminal chlamydospores
Colony salmon color
No growth on rice medium
Microsporum audouinii
Many microconidia: smooth, elongated macroconidia few or absent
Urea agar
Positive in 2 days
Potato dextrose agar
Star-shaped, powdery colonies
Round, grape-like clusters of microconidia on cornmeal agar Few spiral hyphae present in some cultures
Trichophyton mentagrophytes
Negative in 2 days
cornmeal dextrose agar
Red pigment
Tear-shaped microconidia along hyphae on cornmeal agar
Trichophyton rubrum
No red pigment
Balloon-shaped microconidia on cornmeal agar
Trichophyton agar 4
Increased growth
Trichophyton tonsurans
Conidia usually not present, only hyphae seen
Colony violet color
Trichophyton agar 4
Increased growth
Trichophyton violaceum
Colony white and wrinkled
Antler hyphae present
Trichophyton schoenleinii
Colony very slow growing, heaped up, partially submerged in medium
Colony surface smooth without aerial hyphae
Antler hyphae sometimes present
Increased growth at 37°C with production of chains of chlamydospores with septa appearing like fission planes
Trichophyton agar 4
Increased growth
Trichophyton verrucosum

* Schema used for dermatophytes commonly recovered by Mayo Clinic Mycology Laboratory.

walled macroconidia, commonly borne either singly or in clusters of two or three from the tips of short conidiophores (Plate 6.6*C*). Chlamydospores may be numerous in older cultures.

Microconidia are not produced, an important differential feature in separating *E. floccosum* from *Trichophyton* sp and those species of *Microsporum* in which the roughened walls are not prominent (*M. nanum* for example—Plate 6.6*D*).

The gross colony appearance may give some clue to the identification of *E. floccosum*; it characteristically is olive green to khaki-colored, although white or gray-white variants, sometimes surrounded by a dull orange-brown apron, are occasionally encountered (Fig. 4 of Color Plate 3).

*E. flocossum* is commonly recovered from patients with tinea pedis (athlete's foot) or tinea cruris. It can also infect the nails, but does not invade the hair. The fungus is anthropophilic and rarely infects animals.

### Features of *Epidermophyton floccosum*

- Macroconidia are large, club-shaped, multicelled, smooth-walled and borne singly or in clusters of two or three
- Microconidia are not formed
- Colonies are cottony (floccose), initially white but often take on a khaki color with age

### Dermatophyte Genus Identification: Microsporum

Macroconidia are large, have thick, rough walls divided into many cells by transverse septa
Microconidia are relatively few or absent

## THE SPECIES OF MICROSPORUM

### Microsporum canis

The presence of spindle-shaped macroconidia with thick, roughened walls, which have a tendency for their pointed anterior tips to slightly curve to one side is highly suggestive of ***M. canis*** (Plate 6.6*E*). The colonies are granular or fluffy with a feathery border, white to buff and characteristically have a lemon-yellow or yellow-orange apron at the periphery (Fig. 2 of Color Plate 3) that can also be seen when examining the reverse of the colony. Microconidia are usually few in number; however, occasionally large numbers may be seen.

### Features of *Microsporum canis*

- Macroconidia multicelled with roughened walls; spindle-shaped, terminal end sometimes curved
- Microconidia few or usually absent

### Microsporum gypseum

The differentiation of ***M. gypseum*** from *M. canis* is usually easily accomplished. Most strains of *M. gypseum* produce dense aggregates of macroconidia, but many microconidia may sometimes be present. This leads to a much more granular appearance of the colony (Fig. 3 of Color Plate 3). Each macroconidium is broadly spindle-shaped with a rounded rather than a pointed or tapered, curved tip (Plate 6.6*F*). The surface of the colony is granular and cinnamon brown in color rather than the lighter lemon yellow characteristic of most strains of *M. canis*. The appearance of the colonial and microscopic morphology is usually sufficient to make the differential diagnosis. *M. canis* is zoophilic;

*M. gypseum* is geophilic, differences that may have important epidemiological implications in some cases.

**Features of *Microsporum gypseum***

- Macroconidia are multicelled with roughened walls; terminal ends are rounded
- Microconidia may be present, singly or in small clusters
- Colony granular and cinnamon colored

**Microsporum andouinii**

Unfortunately the aglorithm presented in Table 6.4 often does not provide definitive morphological criteria to identify *M. audouinii* because this species usually fails to sporulate. If macroconidia are observed, they are usually bizarre in shape and distorted so that a specific identification is difficult to make (Plate 6.6*G*). Microconidia are virtually never observed; rather, various atypical vegetative forms such as terminal chlamydospores, favic chandeliers, and racquet hyphae are the only clues to the genus (Plate 6.6*H*).

Colonies of *M. audouinii* generally grow more slowly than other members of the genus *Microsporum* (usually 10–21 days), producing a velvety aerial mycelium that is colorless to light tan to gray. The reverse side often appears salmon pink. *M. audouinii* does not grow on rice medium (Appendix II), a helpful characteristic to differentiate *M. canis* which can grow on this medium.

**Features of *Microsporum audouinii***

- Conidia are almost never formed; if present, bizzare and difficult to identify
- Terminal chlamydospores and antler hyphae often present
- Colonies slow growing; no growth on sterile rice grains

It is not uncommon to identify *M. audouinii* by exclusion, although if the features described above are found in a skin or hair culture obtained from a child with tinea capitis, a more definitive identification is possible, particularly if the infected hairs produce an apple-green fluorescence when observed under a Wood's lamp.

**Dermatophyte Genus Identification: Trichophyton**

Many microconidia are present, in grape-like clusters or singly along the hyphae

Macroconidia few or absent, elongated and pencil-shape; thin smooth walls without echinulations and many cells separated by transverse septa

## THE SPECIES OF TRICHOPHYTON

***T. mentagrophytes*** and ***T. rubrum*** are the two *Trichophyton* sp most commonly recovered in clinical laboratories. These two species can be differentiated by observing several characteristics.

### Trichophyton mentagrophytes and Trichophyton rubrum

The colonies may be either fluffy or granular; zoophilic strains of *T. mentagrophytes* recovered from human infections have a tendency to be granular and are associated with inflammatory infections. As the name would indicate, *T. rubrum* tends to produce abundant wine-red, water soluble pigment that diffuses into the agar medium (Fig. 7 of Color Plate 3). However, many strains of *T. mentagrophytes* may also produce a red pigment, although usually much less intense than that produced by *T. rubrum*, particularly when colonies are grown side by side on a marginally nutritious medium such as cornmeal agar.

Both species produce abundant microconidia. When grown on cornmeal agar, *T. mentagrophytes* tends to produce round microconidia arranged in compact, grape-like clusters (*en grappe*) and spiral hyphae are commonly observed (Plate 6.7*A*). Conversely, tear-shaped microconidia of *T. rubrum* tend to be dispersed singly and laterally along the hyphae (*en thyrses*—Plate 6.7*B*). Both species produce elongated, cigar-shaped or pencil-shaped macroconidia with smooth walls (Plate 6.7*C*). Those of *T. mentagrophytes* are connected to the hyphae by a definite narrow attachment; those of *T. rubrum* are attached directly to the hyphae without an identifiable connector.

**Features Differentiating *T. mentagrophytes* from *T. rubrum***

| *T. mentagrophytes* | *T. rubrum* |
|---|---|
| Microconidia spherical, often produced in clusters | Microconidia small, tear-shaped, often borne laterally on the hyphae |
| Macroconidia usually few in number, multicelled, smooth-walled and cigar-shaped, with a definite narrow attachment at the base | Macroconidia usually few in number, multicelled, smooth walled, pencil-shaped and attached directly to hyphae. |
| Spiral hyphae in ⅓ of isolates | |
| Colonies fluffy or granular; production of red pigment usually scant, particularly when grown on cornmeal agar | Colonies generally fluffy; abundant wine-red, water-soluble pigment produced coloring the culture medium |
| Urease produced in 1–3 days. | Urease either not produced or delayed weak reaction |

*T. mentagrophytes* produces urease within 1–3 days after inoculation of Christensen's urea agar, whereas *T. rubrum* is either negative for urease or produces a delayed faint pink color in the culture medium after 7 days of incubation. The definitive characteristic for differentiating the two species can be determined by performing the hair baiting test as follows:

**Hair Baiting Test**

1. Place a filter paper disk (approximately 90 mm in diameter) into the bottom of a standard 100-mm sterile Petri dish, add two drops of yeast extract
2. Add approximately 15 ml of sterile water
3. Place a lock of sterilized child's hair into the water
4. Transfer a portion of the colony to be studied also directly into the water in the Petri dish
5. Replace the lid of the Petri dish and incubate the culture at 25°C (room temperature) for 10–14 days.
6. Microscopic observations of the hairs can be made at regular intervals, usually not fruitful during the initial 10 days. To make an examination, place a

few hairs into a drop of water on a microscope slide, overlay a coverslip and examine under low and high power magnification for the presence of conical-shaped penetrations of the hair shaft.

7. *T. mentagrophytes* has the ability to invade the hair shaft (Plate 6.7*D*); *T. rubrum* grows on the surface but does not penetrate

### Trichophyton tonsurans

***T. tonsurans*** may be suspected if a colony similar to that seen in Figure 5 of Color Plate 3 is observed in a culture obtained from a scalp lesion involved with tinea capitis of the black dot type. The colony grows slowly and is typically buff or brown, wrinkled and suede-like in appearance.

Microscopically *T. tonsurans* can also be suspected if microconidia of various sizes and shapes and with a flattened base are observed, particularly if clavate (club-shaped), or ballooned forms are seen (Plate 6.7*E*). Macroconidia are only rarely seen and when present are bizzare in shape and distorted.

The definitive identification of *T. tonsurans* and other members of the genus *Trichophyton* can be supported by observing the growth patterns on a series of agars with and without added nutrients. Table 6.5 lists the reactions observed in four Trichophyton agars (three additional agars, 5, 6 and 7, which are not commonly used in most laboratories, are also available, as listed in Appendix II). *T. tonsurans* grows maximally in medium supplemented with thiamine (4+ growth in agars 3 and 4), but grows minimally in culture media deficient in thiamine. As mentioned earlier in this chapter, *T. tonsurans* is being recovered with increasing frequency in many large cities in the United States and microbiologists should be alert to the potential presence of this organism.

**Table 6.5.** Growth Patterns of Common Dermatophytes on *Trichophyton* Differential Agars*

| Organism | Casein Basal Agar No. 1 | Casein + Inositol No. 2 | Casein + Inositol + Thiamine No. 3 | Casein + Thiamine No. 4 |
|---|---|---|---|---|
| *T. verrucosum* | | | | |
| 84% | 0 | ± | 4+ | 0 |
| 16% | 0 | 0 | 4+ | 4+ |
| *T. schoenleinii* | 4+ | 4+ | 4+ | 4+ |
| *T. tonsurans* | ±–1+ | 1+ | 4+ | 4+ |
| *T. mentagrophytes* | 4+ | 4+ | 4+ | 4+ |
| *T. rubrum* | 4+ | 4+ | 4+ | 4+ |
| *T. violaceum* | ±–1+ | 0 | 4+ | 4+ |

4+ = rich abundant growth
1+ = submerged growth of approximate 10 mm
± = growth about 2 mm or less
0 = no growth

* Note: Pure cultures from non-vitamin-enriched medium, such as Sabouraud's dextrose agar or Mycosel should be used. Bacterial contaminated cultures should not be used in that many bacteria synthesize vitamins that can invalidate the test. It is also important when inoculating the different agars that the substances contained in one not be carried over to the next tube. Inocula to the nutrition tubes must be very small. Nutritional tests are incubated at 30° C for 1–2 weeks before interpreting.

**Features of *Trichophyton tonsurans***

- Microconidia are often prominent, usually with flattened bases; with age, they tend to become pleomorphic and swollen to elongated resembling "balloon forms."

- Macroconidia are rare and distorted when found
- Colony slow growing and usually brown or tan in color; growth poor or absent in media deficient in thiamine

### Trichophyton verrucosum

***T. verrucosum*** most commonly causes ectothrix hair infections in man, and should be suspected in symptomatic patients who are farm or ranch hands since this organism causes endemic infection in cattle. The colonies grow very slowly (20–30 days usually required) and appear white or gray with a smooth surface. Growth is enhanced in Trichophyton agars 3 and 4 (all strains require thiamine for growth; others also require inositol) or when the culture is incubated at 35°–37°C. *T. verrucosum* may also be suspected grossly when colonies showing a propensity to cut into the agar and produce subsurface growth are observed.

Microscopically, chlamydospores in chains and antler hyphae may be the only forms observed. The chains of chlamydospores along with hyphae fragments may also be present in deep lesions of the skin and may be confused with *B. dermatitidis* or one of the achlorophilic algae (*Fissuricella filamenta*) when microscopically observed in wet mounts prepared from clinical specimens. Conidia may be produced in cultures if Sabouraud's dextrose agar enriched with yeast extract or thiamine is used. Conidia, when present, are borne laterally from hyphae and are relatively large and clavate (Plate 6.7*F*). Macroconidia are rarely produced; however, when seen, are thin-walled, have 3–5 cells, and may show an elongated projection so as to resemble a "rat tail."

#### Features of *Trichophyton verrucosum*

- Conidia usually absent in routine cultures
- Microconidia, when present, are large, clavate and borne laterally from the hyphae
- Macroconidia, rarely present, are smooth-walled, multicelled and often appear as "rat-tail" forms
- Colonies extremely slow growing and often are submerged in the culture medium
- All strains require thiamine for growth; some also require inositol
- Growth enhanced at 37° C

### Trichophyton schoenleinii

***T. schoenleinii*** is slow growing in culture (30 days or more may be required before a mature colony develops). The colony is small, wrinkled, white to light gray in color and has a suede or waxy surface. Microscopically, conidia are not commonly formed; usually only antler hyphae and chlamydospores are observed (Plate 6.7*G*).

*T. schoenleinii,* a cause of the favus type of tinea capitis infection and is endemic in northern Europe, particularly in Scandinavian countries. This fungus is rare the United States except in Appalachia.

#### Features of *Trichophyton schoenleinii*

- Conidia absent
- Antler hyphae and chlamydospores are common
- Colonies slow growing, heaped, wrinkled and smooth to waxy

### Trichophyton violaceum

***T. violaceum*** also causes a favus type of tinea capitis and is endemic in southern Europeans, particularly in ethnic groups living in countries surrounding the Mediterranean sea. This fungus is also very slow growing, requiring a month or more to mature. This fungus can be suspected by observing wrinkled, yeast-like colonies that produce the characteristic violet or deep port wine pigment (Fig. 8 of Color Plate 3).

Growth may be enhanced by using culture media enriched with thiamine, such as Trichophyton agar 3 and or 4.

Microscopically, conidia are absent and swollen cells with cytoplasmic granules and distorted hyphae are the only structures that may be observed (Plate 6.7*H*).

**Features of *Trichophyton violaceum***

- Conidia usually absent
- Swollen hyphae containing cytoplasmic granules may be seen
- Colonies are extremely slow growing, are violet or purple in color and waxy in consistency
- Growth enhanced on media containing thiamine

The colonial and microscopic features of the 10 species of dermatophytes discussed above are summarized in Table 6.6. How much emphasis any given laboratory should place on the recovery and identification of dermatophytes will be dictated by the interest of the laboratory director or supervisor and by the needs of the physicians being served. As a point of reference, Table 6.2 lists the numbers of dermatophytes recovered from clinical specimens submitted to the Mayo Clinic Mycology Laboratory between 1977 and 1984. Notice that *T. rubrum* is recovered about three times more frequently than all other dermatophytes combined, and that *Microsporum* sp is relatively infrequent. It can be assumed that other species of dermatophytes must only rarely be encountered in other laboratories in the United States and therefore should receive the least emphasis in teaching new students. The Suggested Readings cited at the end of this text should be consulted for descriptions and illustrations of the dermatophyte species other than those discussed here.

## THE DIMORPHIC FUNGI

The species of fungi comprising the dimorphic fungi and which cause the deep or systemic mycoses in humans are:

| Fungal Species | Human Mycosis |
|---|---|
| *Blastomyces dermatitidis* | Blastomycosis |
| *Paracoccidioides brasiliensis* | Paracoccidioidomycosis |
| *Histoplasma capsulatum* | Histoplasmosis |
| *Coccidioides immitis* | Coccidioidomycosis |
| *Sporothrix schenckii* | Sporotrichosis |

Dimorphism refers to the propensity of the pathogenic species of fungi listed above to present two growth forms: (a) a mold when grown at 25°–30°C (room or environmental temperature) and (b) a yeast at 37°C

**Table 6.6.** Characteristics of More Commonly Isolated Dermatophytes

| Dermatophyte | Colonial Morphology | Growth Rate | Microscopic Identification |
|---|---|---|---|
| *Microsporum audouinii* | Downy white to salmon pink colony; reverse tan to salmon pink | 2 weeks | Sterile hyphae: terminal chlamydospores, favic chandeliers, and pectinate bodies; macroconidia rarely seen—bizarre shaped if seen; microconidia rare or absent |
| *Microsporum canis* | Colony usually membranous with feathery periphery; center of colony white to buff over orange-yellow; lemon yellow or yellow-orange apron and reverse | 1 week | Thick-walled, spindle-shaped, multiseptate, rough-walled macroconidia some with a curved tip; microconidia rarely seen |
| *Microsporum gypseum* | Cinnamon-colored, powdery colony; reverse light tan | 1 week | Thick-walled rough, elliptical multiseptate macroconidia; microconidia few or absent |
| *Epidermophyton floccosum* | Center of colony tends to be folded and is khaki green; periphery is yellow; reverse yellowish brown with observable folds | 1 week | Macroconidia large, smooth-walled, multiseptate, clavate, and borne singly or in clusters of two or three; microconidia not formed by this species |
| *Trichophyton mentagrophytes* | Different colonial types; white to pinkish, granular and fluffy varieties; occasional light yellow periphery in younger cultures; reverse buff to reddish brown | 7–10 days | Many round to globose microconidia most commonly borne in grape-like clusters or laterally along the hyphae; spiral hyphae in 30% of isolates; macroconidia are thin-walled, smooth, club-shaped, and multiseptate. Numerous or rare depending upon strain |
| *Trichophyton rubrum* | Colonial types vary from white downy to pink granular; rugal folds are common; reverse yellow when colony is young; however, wine red color commonly develops with age | 2 weeks | Microconidia usually teardrop, most commonly borne along sides of the hyphae; macroconidia usually absent, but when present are smooth, thin-walled, and pencil-shaped |
| *Trichophyton tonsurans* | White, tan to yellow or rust, suede-like to powdery; wrinkled with heaped or sunken center; reverse yellow to tan to rust red | 7–14 days | Microconidia are teardrop or club-shaped with flat bottoms; vary in size but usually larger than other dermatophytes; macroconidia rare and balloon forms found when present |
| *Trichophyton schoenleinii* | Irregularly heaped, smooth white to cream colony with radiating grooves; reverse white | 2–3 weeks | Hyphae usually sterile; many antler-type hyphae seen (favic chandeliers) |
| *Trichophyton violaceum* | Port wine to deep violet colony, may be heaped or flat with waxy-glabrous surface; pigment may be lost on subculture | 2–3 weeks | Branched, tortuous hyphae that are sterile; chlamydospores commonly aligned in chains |
| *Trichophyton verrucosum* | Glabrous to velvety white colonies; rare strains produce yellow-brown color; rugal folds with tendency to sink into agar surface | 2–3 weeks | Microconidia rare; large and tear-drop when seen. Macroconidia extremely rare, but form characteristic "rat-tail" types when seen; many chlamydospores seen in chains, particularly when colony is incubated at 37° C |

(human body temperature). Figure 1 of Color Plate 5 illustrates a dimorphic colony including yeast and mold areas.

In the natural environment, the dimorphic fungi exist in soil, particularly soil enriched with bird manure and in dust in the mold form. In the United States, *B. dermatitidis* and *H. capsulatum* are endemic in the **Mississippi, Missouri, and Ohio river valleys** and their tributaries where the soil is moist and relatively cool. Birds serve as mechanical carriers of *H. capsulatum* and conidia may be in high concentrations in natural areas of roosting where dried excreta accumulate. *P. brasiliensis* is not endemic in the United States; rather, as the species name indicates, is found primarily in South America. *C. immitis* is endemic in the **hot, dry sands of the arid Southwestern United States** (particularly in the **San Joaquin Valley**), in Mexico, Central and South America. The arthroconidia germinate maximally about ½ inch below the desert sand where the soil temperature reaches an optimum 160°F. These conidia easily become airborne in dust storms. *S. schenckii* is virtually **worldwide** in distribution, although found sparingly in the Rocky Mountains. Its habitat is plants widely distributed in nature.

The mold form of the dimorphic fungi is the infective form for humans. The modes of infection are either through inhalation of air or dust contaminated with conidia or hyphal fragments (although this route of infection has not been definitely demonstrated for *B. dermatitidis*) or by direct inoculation of broken skin or mucous membranes with soil or vegetative matter. Direct inoculation is particularly apropos for *S. schenckii*, which colloquially is known as the rose gardener's fungus because this occupation puts one at high risk of pricking the fingers and hands with conidia-contaminated thorns.

The yeast or spherule form of *C. immitis* is responsible for the tissue reaction in cases of human infection and the untoward complications. These forms are not infective; rather, the organisms must convert back to their mold form by reentry into the environment where conidiogenesis can take place. Thus, laboratory workers have minimal risk of acquiring infection from handling surgical or autopsy tissue obtained from cases of systemic mycoses, but are under considerable risk when working with the mold forms of these organisms that are recovered in culture.

The avidity with which clinical microbiologists attempt to recover species of dimorphic fungi from clinical specimens to some extent depends upon the locale of the laboratory in which they are employed. A concerted effort must be made to recover these fungi from patients with suspicious symptomatology who live in endemic regions of the country, or from patients who give a history of recent travel into such regions. Culture techniques, therefore, will vary from laboratory to laboratory depending upon the prevalence of disease.

It is generally recommended that primary cultures for the recovery of fungi be incubated at 30°C. Thus, the dimorphic fungi will be recovered only in their mold forms and must be converted to the yeast form before a definitive identification can be confirmed. The conversion of dimorphic molds (except *C. immitis*) can be accomplished with some difficulty as follows:

**Yeast Conversion of Dimorphic Molds**

1. Transfer a rather large inoculum of the filamentous culture onto the surface of a fresh, moist slant of brain-heart infusion agar containing sheep blood. If *B. dermatitidis* is suspected, a plate of cottonseed conversion medium (see Appendix II)

can also be set up to accomplish a more rapid conversion
2. Add a small amount of brain-heart infusion broth or a few drops of sterile water to provide moisture to the culture during incubation
3. Tighten the screw cap, then loosen just slightly to allow the culture to "breathe"
4. Incubate cultures at 37°C for several days, observing for the appearance of yeast-like portions of the colony. It may be necessary to make subsequent subcultures of any growth that appears in that more than one passage is often required to accomplish full yeast conversion
5. Under a safety hood and using appropriate techniques, prepare mounts of any growth and observe for the characteristic yeast forms. If conversion has taken place, report out the appropriate species depending upon the morphology observed

Since the conversion of molds to the corresponding yeast or spherule forms is technically cumbersome to perform and long delays are often experienced before a definitive identification can be made, the exoantigen test is recommended. This relatively new technique is being used in many laboratories to make the definitive identification of *B. dermatitidis, H. capsulatum, P. brasiliensis*, and *C. immitis*. The exoantigen test is based on the principle that soluble antigens are produced by these fungi and may be concentrated and subsequently reacted with sera known to contain antibodies directed against the specific antigenic components. Reagents are commercially available which provide virtually all clinical laboratories with the capability of identifying these species of fungi with ease and rapidity. The techique is particularly helpful in the early study of cultures with cobweb or fluffy colonies suspected of being *C. immitis*, some strains of which may appear as early as 3–5 days after inoculation of culture media. Yeast conversion studies must, however, still be performed to identify *S. schenckii* definitively for which an exoantigen method is not currently available. The exoantigen test is performed as follows:

### Exoantigen Test for Identification of Certain Dimorphic Fungi

1. A mature fungus culture on a Sabouraud's dextrose agar slant is covered wih an aqueous merthiolate solution (1:5000 final concentration) and is allowed to react with the culture for 24 hours at 25° C (be sure the surface of the entire colony is covered to render the fungus safe). Filter the aqueous solution through a 0.45-$\mu$m nalgene filter.

2. Five ml of this solution is concentrated to 50× for *H. capsulatum* and *B. dermatitidis* and 5× and 25× for *C. immitis*, using an Amicon Minicon Macrosolute B-15 Concentrator (Amicon Corp., Danvers, MA).

3. The concentrated supernatant is set up by the microimmunodiffusion test in wells adjacent to control antigen and tested against positive control antiserum obtained commercially.

4. The immunodiffusion test is allowed to react for 24 hours at 25° C and then read for lines of identity with the reference bands as shown in the diagram below. Note: *B. dermatitidis* sensitivity may be increased by incubating the immunodiffusion plates at 37° C for 48 hours; however, the band appears sharper at 25° C for 24 hours. It is recommended that any culture suspected of being *B. dermatitidis* be set up and incubated under both conditions. The immunodiffusion set up is illustrated as follows:

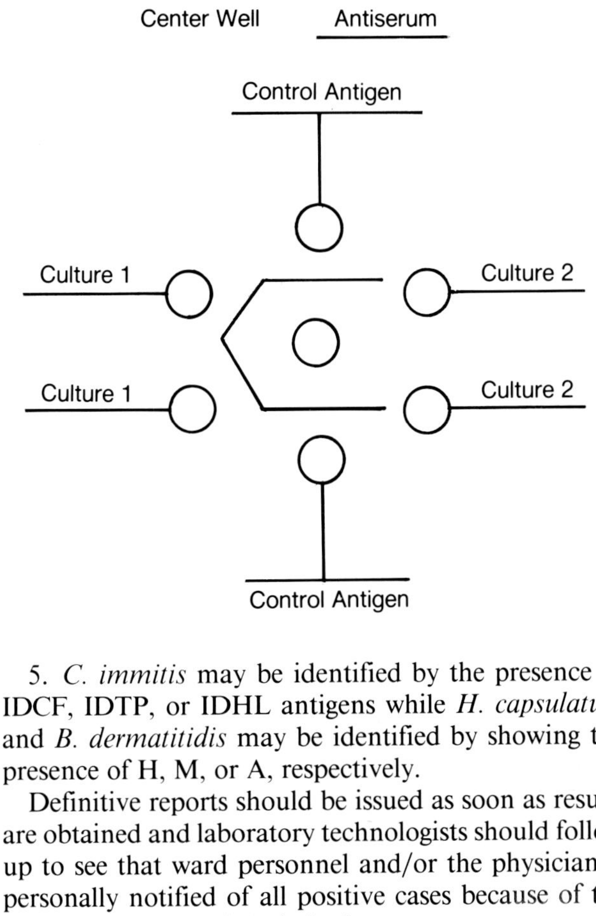


5. *C. immitis* may be identified by the presence of IDCF, IDTP, or IDHL antigens while *H. capsulatum* and *B. dermatitidis* may be identified by showing the presence of H, M, or A, respectively.

Definitive reports should be issued as soon as results are obtained and laboratory technologists should follow up to see that ward personnel and/or the physician is personally notified of all positive cases because of the potential severity of the infection.

### *Blastomyces dermatitidis*

Blastomycosis is a fungal disease that may occur in two general forms: (a) **primary pulmonary** with secondary dissemination to other organs; and (b) **primary cutaneous**, usually limited to one part of the body (33). The primary cutaneous form is a chronic suppurative and granulomatous disease of the dermis. The presence of intraepithelial microabscesses including the characteristic yeast forms of ***B. dermatitidis*** is diagnostic. This form is usually self-limited and does not disseminate. It is incumbent upon the physician to rule out dissemination in all cases of cutaneous or mucous membrane involvement by performing a general physical examination and obtaining X-rays as appropriate and cultures of sputum and urine.

Patients with pulmonary blastomycosis may present initially with dry cough, pleuritic pain, and low grade fever. Dissemination to distant organs, commonly the skin, oral and nasal mucosa, bone and central nervous system may occur.

The primary source of the organism is probably from the soil, although the exact epidemiology of human infections has not been elucidated. The fungus is endemic in the Southeast, Midwest and Southern United States, particularly in the great river valleys.

The laboratory recovery of the mycelial form is facilitated by the use of an enriched medium, such as brain-heart infusion agar containing blood or yeast extract supplements. The mycelial form usually requires 5 days to 4 weeks or longer for primary growth; the yeast form can usually be converted within 5 days after transfer from a mold culture, particularly if cottonseed conversion agar is used (formula for cottonseed agar is in Appendix II).

The definitive diagnosis of blastomycosis may be made by observing in body fluids or in histological sections of infected tissue the characteristic large (8–15 μm) thick-walled, singly budding yeast cells of *B. dermatitidis* in which the daughter cell is attached by a broad base ("broad-based-buds"—Plate 3.1*A* and *B*).

On enriched culture medium, the mycelial form develops initially as a moist, glabrous or waxy-appearing mold that may develop an off-white, delicate silky aerial mycelium, often turning gray or brown with age (Fig. 2 of Color Plate 5). On culture media enriched with blood, the colony assumes more of a waxy, yeast-like appearance. "Tufts" of hyphae often project upward from the colonies (Fig. 3 of Color Plate 5). *B. dermatitidis* is usually considered to be a slow-growing organism, requiring 14 days to 4 weeks to grow; however, specimens containing a large inoculum can show evidence of visible growth by 5 days.

Conversion to the yeast form or the demonstration of exoantigen must be accomplished before a definitive identification can be made.

Microscopically, the hyphae of the mold form are septate and delicate, measuring 1–2 μm in diameter (Plate 6.8*A*). A parallel arrangement of hyphae may be seen in colonies grown on medium containing blood (Plate 6.8*B*). The diagnostic forms are single, pyriform conidia produced on short to long conidiophores, giving the overall appearance of "**lollypops**" (Plate 6.8*C* and *D*). Production of conidia is sparse in some cultures and are usually not produced on a medium containing blood.

In the yeast form, the colonies appear tan or cream in color, are waxy and quite wrinkled (Fig. 5 of Color Plate 5). Similar colonies may also be produced on mycobacterial culture medium incubated at 37° C. Microscopically, large, thick-walled yeast cells in which a single bud is attached to the parent cell by a thick "collar" are seen (Plate 6.8*E*). During the conversion process, abortive hyphal forms and immature cells with rudimentary single buds may be seen (Plate 6.8*F*).

## DIFFERENTIAL CONSIDERATIONS

The mold form of *B. dermatitidis* must be differentiated from the saprobe *Chrysosporium*, from *P. boydii* and from *Trichophyton* sp. All of these fungi produce single conidia either directly from the hyphae or from delicate conidiophores simulating lollypops. *B. dermatitidis* can be differentiated by the usually slow growth rate (usually 14 or more days), by the ability to grow on medium containing cycloheximide and the ability to convert to a yeast form when incubated at 37° C on enriched medium. The production of the specific immunodiffusion band in the exoantigen test also confirms the identification.

The yeast form of *B. dermatitidis* should not be confused with the yeasts of *C. neoformans* (which are usually encapsulated) or of *H. capsulatum*, which are considerably smaller (2–3 μm). Both of these yeasts produce buds with a delicate pinched-off attachment (rather than a broad base) to the mother cell. Immature spherules of *C. immitis*, which have not yet developed endospores and may be as small as 10–15 μm, may be confused with the yeast forms of *B. dermatitidis* when lying adjacent to each other; however, *C. immitis* spherules never produce buds.

### Features of *Blastomyces dermatitidis*

- Conidia of mold form are small, round to pyriform on the tips of long or short conidiophores
- Yeast forms at 37° C are large, have broad-based buds, with a cytoplasm appearing retracted from the cell walls
- Colonies are slow growing, usually 14 or more days (some strains may grow as quickly as 5 days)
- Specific A-band seen in the exoantigen test

## SEROLOGICAL DIAGNOSIS

The complement fixation test has been the most widely used test for the serological diagnosis of blastomycosis; however, its value is limited. Less than 25% of culturally proven cases of blastomycosis are detected using this method. Titers of 1:8 to 1:16 are suggestive of active disease and titers of 1:32 may occur in some patients with blastomycosis. Cross-reactions occur in sera of patients with histoplasmosis and titers may be equivalent to those achieved using the *B. dermatitidis* antigen, making interpretations sometimes difficult.

The immunodiffusion test has been reported to detect at least 80% of active cases of blastomycosis. The presence of a precipitin band is indicative of active disease; however, data in the medical literature are conflicting regarding the sensitivity of this test. Nevertheless, in cases where the diagnosis is clinically highly suspicious, immunodiffusion test results may be useful.

### *Paracoccidioides brasiliensis*

Paracoccidioidomycosis (South American blastomycosis) is almost exclusively limited to South and Central America and is caused by the dimorphic fungus ***P. brasiliensis***. The disease is primarily pulmonary in origin, where it can present as a micronodular infiltration simulating tuberculosis. The disease also frequently involves the mouth, nasal cavities, tonsils, or larynx where progressive suppurative and granulomatous ulcers are produced. Dissemination to deeper organs may also occur.

The mycelial form grows in culture very slowly (21–28 days), although there are no special growth requirements. The colony surface is smooth and covered by a low aerial mycelium which is brown or off-white. The colony is often heaped with formation of a central crater (Fig. 7 of Color Plate 5).

Microscopically, the mold form is characterized by hyphae measuring 1–2 μm in diameter. The mycelium tends to be sterile and numerous chlamydospores may be seen. When present, globose or pyriform conidia similar to those produced by *B. dermatitidis* are present (Plate 6.8*G*).

The colony morphology of the yeast form is similar to that described for *B. dermatitidis*. Microscopically, the yeast forms are typically large (10–40 μm) in diameter, although in some instances cells are within the size range of *H. capsulatum*. The mother cell is similar to that produced by *B. dermatitidis*; the difference, however, is that multiple small or large daughter buds are produced around the circumference of *P. brasiliensis* yeast cells, often simulating a mariner's wheel (Plate 3.1*D* and Plate 6.8*H*). Small budding yeast cells may resemble *H. capsulatum*.

#### Features of *Paracoccidioides brasiliensis*

- Characteristic conidia often not formed; rather, numerous chlamydospores and twisted hyphae may predominate; when present, conidia are similar to those of *B. dermatitidis*
- Large (10–40 μm) yeast cells show multiple buds, simulating a "mariner's wheel"

#### SEROLOGICAL DIAGNOSIS

The complement fixation test is positive in 80–90% of cases with active infection, with titers of 1:8 or greater. Cross-reactions occur with other fungal antigens, but titers are usually lower than those obtained with the paracoccidioides antigen. The immunodiffusion test is positive in approximately 94% of patients with active infection. Precipitin bands 1, 2, and 3 may be present; band 1 is most commonly seen.

### *Histoplasma capsulatum*

Histoplasmosis is a chronic granulomatous disease initially involving the lungs, and in a small percentage of patients, the disease may progress to a disseminated form with general involvement of the reticuloendothelial system and various visceral organs. Patients with disseminated disease experience hepatosplenomegaly, lymphadenopathy, fever, anemia, and weight loss.

In most infections, an acute bronchopneumonia may occur shortly after inhalation of conidia; however, within 1–2 weeks, spontaneous healing occurs. In rare cases, a chronic pulmonary cavitary form may evolve, at times healing with the formation of one or more fibrous nodules that may undergo calcification (onion-skin granulomas). Histologically, the lesions in the lungs and other tissues are initially suppurative, progressing to a caseating, granulomatous reaction simulating tuberculosis.

The yeast form consists of small, 2–5 μm budding yeast cells that show narrow attachments to the mother cell. In the tissues, the yeasts are intracellular within large fixed macrophages of the reticuloendothelial system or lying free in the extracellular spaces. In stained sections, the yeast cells appear to be surrounded by a capsule (Plate 3.2*E* and *F*). This appearance accounts for the species name, although a true capsule has not been demonstrated. The pseudocapsule effect is thought to be secondary to a shrinking fixation artifact, with the yeast body pulling away from the other membrane during processing.

***Histoplasma capsulatum*** is usually considered to be a slow-growing mold at room temperature or at 30° C, requiring 2–4 weeks or more for colonies to appear; however, growth can occasionally be seen in less than one week if a large concentration of organisms is present in the specimen. The colonies are white, wrinkled, moist and sometimes initially yeast-like, particularly in culture media supplemented with blood. A silky aerial mycelium develops as the colony matures, in time turning gray or brown (Fig. 2 of Color Plate 5). A prickly stage similar to that of *B. dermatitidis* is common (Fig. 3 of Color Plate 5). Variations in the appearance of cultures may occur, varying with the strain of the isolate, the medium used for culture and the environmental conditions during incubation.

Microscopically the mold form presents delicate 1–2 μm in diameter septate hyphae (Plate 6.9*A*) with rope-like aggregates seen in cultures grown on blood-containing medium. Small round to teardrop microconidia on short, lateral branches along the hyphae may be seen in occasional cultures (Plate 6.9*B* and *C*). More commonly, large 8–14 μm in diameter, spherical or pyriform macroconidia are observed, initially with smooth outer walls (Plate 6.9*D*). On prolonged incubation, the macroconidia become roughened or tuberculate, producing the classical diagnostic form (Plate 6.9*E* and *F*). In primary cultures, macroconidia may align in chains (Plate 6.9*G*). Some cultures fail to sporulate; if *H. capsulatum* or *B. dermatitidis* are suspected, an exoantigen test should be performed.

Conversion to the yeast form may be difficult and may require several transfers at three-day intervals. Microscopic examination of mounts prepared from converted yeast colonies reveals 2–5 μm oval, budding yeast cells, similar to the intracellular forms seen in tissue sections (Plate 6.9*H*). Conversion to the yeast form is often difficult and the exoantigen test is recommended to provide a definitive identification.

## DIFFERENTIAL CONSIDERATIONS

*H. capsulatum* mold cultures must be distinguished from saprobic *Sepedonium* sp, which also produces large tuberculate macroconidia. The differential features have been covered in the discussion on *B. dermatitidis* on page 110. The yeast forms can be confused with those of *Candida* (*Torulopsis*) *glabrata*, nonencapsulated variants of *C. neoformans* or the released endospores of *C. immitis*, all of which are within the size range of *H. capsulatum*. Recovery of the fungus in the mold form may be required in most instances to make the differential identification.

**Features of *Histoplasma capsulatum***

- Macroconidia are spherical to pyriform with a roughened, tuberculate surface of the cell wall
- Single-celled, pyriform microconidia may be seen (Some isolates fail to sporulate)
- Yeast forms are small (2–4 μm) at 37° C, with buds attached by a narrow neck
- Colonies slow growing (usually 2 weeks or more); growth may appear within 1 week if inoculum is heavy
- H and/or M bands present in exoantigen test

### SEROLOGICAL DIAGNOSIS

The complement fixation test is most widely used and utilizes both yeast and mycelial antigens. A titer of 1:8 or greater is considered as positive and clinically significant; titers within the 1:8 or 1:16 range may be clinically insignificant in some cases. A complement fixation titer of 1:32 or greater, or a rise in titer during the course of infection is highly significant. The complement fixation test becomes positive within 10–20 days after primary exposure and approximately 70% of patients will develop titers within this time. The mycelial antigen appears to be more often positive during chronic histoplasmosis while the yeast form antigen is more sensitive and more frequently detects acute infection. Rising complement fixation titers are demonstrable in greater than 95% of patients with primary infection while highly elevated titers may not be as common in patients with severe disseminated infection.

Precipitating antibody detected by immunodiffusion or counter-immunoelectrophoresis has been shown to be useful in the diagnosis of histoplasmosis. Two precipitin bands, designated H and M, may be present during active infection. The M band appears early in approximately 80% of patients with active disease and persists after recovery. It may appear after a recent histoplasmin skin test in patients having prior contact with *H. capsulatum*. The presence of H and M bands coincidentally is highly specific for active histoplasmosis, seen in 25% of patients with disseminated disease. The H band rarely appears alone and the significance of its singular presence is unclear. Negative immunodiffusion tests do not rule out active infection and all positive and negative results should be confirmed by performing the complement fixation test.

### *Coccidioides immitis*

Coccidioidomycosis is endemic in the desert Southwestern United States and in the San Joaquin valley of California where over 95% of residents have positive coccidioidin skin tests. The primary disease is pulmonary, secondary to inhalation of airborne arthroconidia, and in the vast majority of cases is a short-lived, febrile, influenza-like syndrome that resolves without sequelae. Cough, pleuritic chest pain, night sweats and anorexia are common symptoms that clear within two weeks to one month. In a few patients, a more chronic course may occur, characterized by low grade fever, weight loss, and malaise. Pulmonary cavitation may occur; or, more commonly, a peripheral residual pulmonary coin lesion is seen on a chest roentgenogram. These are usually surgically removed and often yield the organism in culture. Only in rare cases does the disease become disseminated, beginning with diffuse pneumonia with secondary dissemination to the skin, subcutaneous tissue, bones, and various visceral organs. Progression may be slow; or, in rare instances, rapidly fatal.

Primary cutaneous coccidioidomycosis is rare and the primary lesion often heals without further evidence of disease. We recently encountered a case of primary coccidioidomycosis of the great toe in a Mexican dancer. Progressive osteomyelitis occurred in this particular patient and a surgical amputation of the toe was necessary. There was no evidence of disseminated disease.

The tissue reaction in coccidioidomycosis is usually granulomatous, and often multinucleated giant cells of the Langhan's type and caseation necrosis are present, closely simulating the lesion of tuberculosis. The histological diagnosis can be confirmed by observing the presence of characteristic spherules ranging from 10 to over 60 $\mu$m in diameter, readily detected either in routine H & E stained sections or in PAS or GMS fungal stains (Plate 3.2*C*). Immature forms may be devoid of endospores; however, usually larger endospore-filled spherules are present. In rare instances, only released endospores may be seen in the tissues, a picture that may simulate the yeast forms of *H. capsulatum*; however, these forms never exhibit budding. In rare instances, particularly in the presence of cavitation, rudimentary hyphae may be seen in tissue sections (Plate 3.2*D*).

Cultures of ***C. immitis*** represent a major biohazard to laboratory personnel and appropriate precautions using biosafety techniques and equipment must be followed when examining these cultures. One initial clue that an unknown mold culture may be *C. immitis* is the appearance of a delicate, cobweb growth on enriched media such as brain-heart infusion or inhibitory mold agar within 3 to 21 days after inoculation of the specimen (Fig. 4 of Color Plate 5). On blood agar, a greenish discoloration of the colony may be seen due to leaching of bilirubin byproducts from red blood cells.

Direct microscopic mounts should be prepared in a Class II biological safety cabinet, first taking the precaution of adding a few milliliters of sterile distilled water to the agar surface to wet down the mycelium. The use of a surgical mask to cover the mouth and nose is highly recommended during preparation of this mount. A small portion of the mycelium can be transferred to a drop of lactophenol-aniline blue stain on a microscope slide and a coverslip applied.

Microscopically, early cultures may show only septate hyphae, often branching at right angles with racquet forms often present (Plate 6.10*A*). As the culture ages, the hyphae become dissociated into arthroconidia, characteristically barrel-shaped and separated one from the other by clear-staining, nonviable cells (Plate 6.10*B*). The arthroconidia are wider than the hyphae. Rectangular hyphae may be seen in those cultures that fail to sporulate. When present, the alternate staining appearance of the arthroconidia is an important diagnostic feature.

*C. immitis* cannot be easily converted to the spherule form in the laboratory. It is recommended that the exoantigen test be performed.

## DIFFERENTIAL CONSIDERATIONS

Immature spherules devoid of endospores lying adjacent may be confused with nonbudding yeast forms of *B. dermatitidis*. If the differential identification cannot be made from direct microscopic examination of material in which spherules are seen, a small portion of the specimen or a drop of water extract can be placed on a microscope slide, a coverslip applied and the slide placed in a humidity chamber and incubated at 37° C for 24 hours. Alternatively the coverslip should be sealed with a vaseline-paraffin mixture to prevent drying and the mount incubated at 37° C for 24 hours. Observe for the presence of multiple germinating tubes eminating from the spherules indicative of *C. immitis* spherules (Plate 6.10*C*). If the forms were yeast cells of *B. dermatitidis*, only single budding, broad-based forms will be seen.

Some members of the ***Gymnoascaceae*** are naturally occurring fungi in the soil that microscopically simulate *C. immitis*. These saprobic molds also produce alternate-staining arthroconidia that tend to be more rectangular, although barrel-shape forms may occasionally be seen (Plate 6.10*D*). The colonies also can be difficult to differentiate from *C. immitis* and exoantigen studies are often required to make an identification. Bands of identity with IDCF, HL, or TP antigens will be present.

***Geotrichum*** sp and ***Trichosporon*** sp also can produce delicate hyphae that break into arthroconidia. The gross colonies usually resemble yeasts; however, cob-web-like aerial hyphae simulating those of *C. immitis* may be seen in older cultures. Microscopically, the arthroconidia of both *Geotrichum* sp and *Trichosporon* sp produce evenly-staining arthroconidia (Plate 6.10*E*), a valuable differential clue to distinguish from the alternate staining arthroconidia of *C. immitis*. The arthroconidia of *Geotrichum* sp characteristically germinate from one corner of the mother cell, forming "hockey stick" cells; those of *Trichosporon* may exhibit blastoconidia but they are usually few in number. Many species of *Trichosporon* are urease positive, a helpful biochemical differentiation from *Geotrichum* sp which are negative.

**Features of *Coccidioides immitis***

- Chains of alternate arthroconidia, some barrel-shaped, some retangular
- Racquet hyphae sometimes seen
- Spherules are large (15–60 μm) and when mature contain nonbudding 2–4 μm endospores
- Colonies cobweb-like, with fluffy areas alternating with areas adherent to the agar surface
- IDCF, TP, or HL antigens present in exoantigen test

### SEROLOGICAL DIAGNOSIS

The complement-fixation test is helpful, with titers of 1:2 to 1:4 suggesting active infection; titers of 1:16 or greater usually indicate active infection. These antibodies are first detected 1–3 weeks after onset of symptoms of primary infection in 75% of cases. Serum IgG complement-fixing antibodies occur later in the disease and may persist for 6 or 8 months or longer. Within 3 months after the onset of infection, 50–90% of patients with symptomatic primary infection have IgG antibodies. Cross-reactions occur in sera of patients having histoplasmosis and blastomycosis; however, titers are usually lower than those occurring with the coccioidin antigen.

The immunodiffusion test is equivalent to the complement-fixation test but is more sensitive for detecting early infection if serum is concentrated 10× before testing. A latex agglutination test appears to be the most sensitive for the detection of precipitating antibody. However, a 6–10% false positive rate and a lack of ability to quantitate antibody are deterring factors for the use of this test.

#### *Sporothrix schenckii*

***Sporothrix schenckii***, a naturally occurring fungus that is found on hay, grasses, rose thorns and other plant material, has a worldwide distribution. Primary human infections of the skin occur most frequently following a puncture wound with an infected thorn; for example with rose thorns, the reason why the disease has also been called "rose gardeners disease."

The primary skin lesion begins as a small nonhealing ulcer, commonly of the index finger or back of the hand. These ulcers can be chancriform in appearance, simulating the chancre of primary syphilis. The lymphatics draining the site of primary involvement may become infected producing a string of secondary elevated subcutaneous nodules that may break down and ulcerate, involving the inner aspects of the arm or leg. Although most cases of sporotrichosis remain localized to the skin and subcutaneous tissues, cases of pulmonary disease have been reported. Only rarely does the infection become disseminated to other organs.

Direct examination of suppurative material or examination of histological sections of tissue biopsies are usually of little diagnostic value because it is difficult to demonstrate the characteristic tiny, cigar-shaped yeast forms, even with special stains. The tissue reaction is usually granulomatous with superimposed suppuration and focal abscess formation. A clue to the diagnosis may come by demonstrating asteroid bodies (Plate 3.3*H*) within the inflammatory infiltrate, although these structures are often difficult to find in routine tissue sections, usually requiring multiple serial sections before one is seen. A typical asteroid body consists of a central rounded or oval nidus surrounding a degenerated yeast cell that in most cases is inconspicuous, from which concentrically radiates an amorphous eosinophilic material that is thought to represent the condensation products of antigen-antibody complexes against the degenerating debris of spent inflammatory cells. In animal tissue, the typical 2–4 μm polymorphous spherical, oval or cigar-shaped budding yeast cells can be more readily identified (Plate 3.3*F* and *G*).

The mycelial form is optimally recovered on most fungal culture media at 30° C. Yeast-like colonies develop within 3–5 days with a white to cream color (Fig. 8 of Color Plate 5). With age, the colony tends to become wrinkled, may develop a delicate hair-like aerial mycelium, and turns dark brown or black in color and leathery in consistency.

Microscopically, the mold-form hyphae are 1–2 μm in diameter, are usually branched and septate and produce delicate conidiophores arising at right angles from the hyphae. Small 1–2 μm pyriform conidia are borne in clusters at the tips of the conidiophores, arranged in flowerettes (Plate 6.10*F*). Each conidium is attached to the conidiophore by an individual, delicate thread-like structure that may require oil immersion and sharp focus to see. It is from this structure that the genus name *Sporothrix* is derived. As the culture ages, slightly dark pigmented conidia may also be borne along the sides of the hyphae in a "sleeve" arrangement (Plate 6.10*G*), simulating the microconidia produced by *T. rubrum*.

Upon incubation of the culture at 37° C, the colony appears soft, cream-colored to white and distinctly yeast-like. Microscopically, singly or multiple budding, spherical, oval or cigar-shaped yeast cells are observed (Plate 6.10*H*). Conversion from the mold to the yeast form is not difficult to accomplish, usually occurring within 1–5 days following transfer to the culture to a medium containing blood.

### DIFFERENTIAL CONSIDERATIONS

The flowerette arrangement of the conidia and the hair-like attachments to delicate conidiophores are features unique to *S. schenckii*. The saprobe *Acremonium* also produces delicate conidiophores and spherical or elongated conidia (Plate 6.5*F* and *G*); however, instead of being arranged in flowerettes, they tend to cluster into a cerebriform-appearing mass and the hair-like attachments characteristic of *S. schenckii* are not seen. Other saprobes, such as *Trichothecium* sp, also produce flowerettes; however, hair-like attachments are absent and the conidia are larger. If only the sleeve-like arrangement of conidia is seen, *S. schenckii* must be differentiated from *T. rubrum*. The saprobes also cannot be converted to yeast forms at 37° C.

When yeast forms are identified, they may be difficult to distinguish from other small yeasts such as those of *H. capsulatum* or nonencapsulated *C. neoformans* which are of a similar size. *S. schenckii* yeast cells tend to be cigar-shaped. In making a laboratory identification, it is helpful to know the clinical history of the patient and the diagnostic impression of the physician.

**Features of *Sporothrix schenckii***

- Hyphae are delicate
- Conidia are produced in a flowerette arrangement at the tip of single conidiophores
- Conidia are connected by a thread-like attachment
- Single and slightly pigmented conidia may be seen along the hyphae in older cultures
- Yeast forms are characteristically cigar-shaped; when round, they may simulate yeast cells of *H. capsulatum*
- Mold to yeast form-conversion is easily accomplished within 12–48 hours at 37° C

### SEROLOGICAL DIAGNOSIS

A whole-cell yeast agglutination test is useful for the serological diagnosis of extracutaneous sporotrichosis and titers of 1:80 or greater are significant. However, there is little need for serological tests for the diagnosis of lymphocutaneous infections.

## SUMMARY

Table 6.7 is a succinct summary of the morphological characteristics and serological test results useful for the laboratory diagnosis of the fungal species causing opportunistic and deep mycotic infections in man. Although the information has been covered in some detail in this chapter, Table 6.7 should be helpful to students who often must assimilate a large amount of information in a short time, and to microbiologists who may wish to quickly recall key items.

**Table 6.7.** Characteristics of Pathogenic Fungi and Serological Tests Useful in Making a Laboratory Diagnosis

<table>
<tr><th rowspan="2">Fungal Species</th><th rowspan="2">Growth Rate (Days)</th><th colspan="2">Cultural Characteristics at 30° C</th><th colspan="2">Microscopic Morphologic Features</th></tr>
<tr><th>Blood Enriched Medium</th><th>Nonblood Enriched Medium</th><th>Blood Enriched Medium</th><th>Nonblood Enriched Medium</th></tr>
<tr><td>Blastomyces dermatitidis</td><td>5–30</td><td>Colonies are cream to tan, soft, moist, wrinkled, waxy, flat to heaped and yeast-like<br>"Tufts" of hyphae often project from colony surface</td><td>Colonies are white to cream to tan, some with drops of exudate present, fluffy to glabrous and adherent to agar surface</td><td>Hyphae 1–2 μm in diameter; some aggregate in rope-like clusters; sporulation rare</td><td>Hyphae 1–2 μm in diameter; single pyriform conidia are produced on short to long conidiophores; some cultures produce few conidia</td></tr>
<tr><td>Cryptococcus neoformans</td><td>3–10</td><td>Colonies are usually dome-shaped, dry, cream to tan and smaller than those on media lacking blood enrichment</td><td>Colonies are dry to mucoid and shiny, dome-shaped, smooth, cream to tan in color; some isolates appear golden to orange on some media, i.e., inhibitory mold agar</td><td colspan="2">Cells are usually spherical, vary in size and may be encapsulated; cells may have more than one "pinched-off" bud present on parent cell</td></tr>
<tr><td>Histoplasma capsulatum</td><td>5–45</td><td>Colonies are heaped, moist, wrinkled, yeast-like, soft and cream, tan or pink in color<br>"Tufts" of hyphae often project upward from surface of colonies</td><td>Colonies are white, cream, tan or gray; fluffy to glabrous; some colonies appear yeast-like and adhere to the agar surface; many variations in colonial morphology occur</td><td>Hyphae 1–2 μm in diameter; some aggregated in rope-like clusters; sporulation is rare</td><td>Smooth-walled macroconidia predominate in young cultures, later become tuberculate with age; may be pyriform or spherical; some cultures produce tiny microconidia</td></tr>
<tr><td>Paracoccidioides brasiliensis</td><td>21–28</td><td colspan="2">Colonies heaped, wrinkled, moist and yeast-like with age; colonies may become covered with short aerial mycelium and turn brown</td><td colspan="2">Hyphae 1–2 μm in diameter; some isolates produce conidia similar to those of B. dermatitidis; chlamydospores may be numerous and multiple budding yeast cells 10–25 μm in diameter may be present</td></tr>
<tr><td>Candida albicans and Candida species</td><td>2–4</td><td colspan="2">Colonies vary in morphology; usually white to tan, shiny to dull, flat to heaped, smooth to wrinkled and moist to dry; some colonies produce pseudohyphal fringes at the periphery; colonies growing on blood enriched medium tend to be somewhat smaller and drier than those growing on nonblood-enriched media</td><td colspan="2">Most species produce either blastoconidia, pseudohyphae, or true hyphae; chlamydospores are produced by C. albicans and by certain strains of C. tropicalis</td></tr>
<tr><td>Mucor species and other Zygomycetes</td><td>1–3</td><td colspan="2">Colonies are extremely fast growing, wooly, cover entire agar surface without a margin, gray to brown to gray-black with age</td><td colspan="2">1. Rhizopus species: Rhizoids are produced at the base of a sporangiophore<br>2. Mucor species: No rhizoids are present</td></tr>
<tr><td>Aspergillus fumigatus<br>Aspergillus flavus<br>Aspergillus niger<br>Aspergillus species</td><td>3–5</td><td colspan="2">Colonies of A. fumigatus are usually blue-green to gray-green, while those of A. flavus and A. niger are yellow-green and black, respectively; other Aspergillus sp exhibit a wide variety of colors; blood enrichment usually has little effect on the colonial morphological features</td><td colspan="2">A. fumigatus: uniseriate heads with phialides covering upper one-half to two-thirds of the vesicle<br>A. flavus: uniseriate, biseriate, or both; phialides covering entire surface of spherical vesicle<br>A. niger: uniseriate with phialides covering entire surface of vesicle; conidia jet black</td></tr>
<tr><td>Coccidioides immitis</td><td>3–21</td><td>Colonies white, fluffy to cobweb-like. Green discoloration may occur from extraction of blood pigments from medium; some isolates are yeast-like and heaped, wrinkled and membranous</td><td>Colonies are fluffy, white but may pigmented gray, orange, brown or yellow; mycelium is adherent to portions of the agar surface</td><td colspan="2">Chains of alternate, barrel-shaped arthroconidia are characteristic; some arthroconidia may be elongated and rectangular; hyphae are small and often arranged in rope-like strands; racquet forms may be seen in younger cultures</td></tr>
</table>

**Table 6.7**—*Continued*

| Screening Tests | Microscopic Morphological Features of Tissue Forms | Confirmatory Tests For Identification | Serological Tests Available |
|---|---|---|---|
| Not available | 8–15 μm broad-based budding cells with double walls are seen; cytoplasmic granulation is often obvious | 1. Broad-based budding cells may be seen after in-vitro conversion on cottonseed agar<br>2. Exoantigen test positive with A or B bands (70% exhibit A bands) | 1. The complement fixation test is positive in only 25% of cases<br>2. The immunodiffusion test is purported to be positive in 80% of active cases; studies performed at the Mayo Clinic show a lower incidence of positive results |
| 1. Urease production<br>2. Phenoloxidase production | 2–15 μm single or multiply budding, spherical cells that vary in size are seen; cells usually encapsulated; some strains are not | 1. Carbohydrate utilization<br>2. Nitrate reduction test<br>3. Pigment on niger seed agar | Serological tests for cryptococcal antigen have high sensitivity for CSF; tests on blood samples positive in less than 30% of cases with pulmonary infection |
| Not available | 2–5 μm in diameter pseudoencapsulated yeast cells seen intracellularly within mononuclear histiocytes | Exoantigen test is positive with H, M, or H & M bands. | 1. Complement fixation test is helpful with titers of 1:8 suggestive of infection; titers of ≥ 1:32 indicates active infection<br>2. Immunodiffusion—H & M bands both present suggest active infection; M band alone may indicate recent skin test; or, early or late infection |
| Not available | 10–25 μm multiply budding yeast cells (buds 1–2 μm) resembling "mariner's wheel"; buds attached by narrow necks | Exoantigen test positive for bands 1, 2 or 3 | 1. Complement fixation test positive in 80–90% of active cases.<br>2. Immunodiffusion test positive in 94% of cases; bands 1, 2 & 3 may be present; band 3 most common |
| Germ tube production by *Candida albicans* | Blastoconidia, 2–5 μm in diameter and pseudohyphae may be present | 1. Germ tube production for *Candida albicans*<br>2. Carbohydrate utilization<br>3. Carbohydrate fermentation | Immunodiffusion test is available; however, conflicting data on the usefulness of the test makes interpretation of results difficult |
| Not available | Large (10–30 μm) ribbon-like, twisted, often distorted pieces of aseptate hyphae (septa may occasionally be seen) | Identification is based on characteristic morphological features | Test are currently not available |
| Not available | Septate hyphae, 5–10 μm in diameter, with parallel walls exhibiting dichotomous branching | Identification is based on microscopic morphology and colonial characteristics; *A. fumigatus* can tolerate temperatures up to 45° C | Immunodiffusion test: precipitating antibodies in 50–70% of allergic bronchopulmonary and 90% of fungal ball cases; usefulness in cases of disseminated disease is still questionable; research studies show some promise |
| Not available | Round spherules, 30–60 μm in diameter containing 2–5 μm spherical endospores are characteristic | Exoantigen test is positive for HL, F, and TP bands. | 1. Complement fixation is helpful; titers of 1:2–1:4 suggestive and 1:16 or greater diagnostic of active infection. Some cross reactivity of serum in cases of blastomycosis and histoplasmosis but titers are lower<br>2. Immunodiffusion tests equivalent but more sensitive for detecting early infection if serum is concentrated 10× before testing |

# PLATE 6.1

## LABORATORY IDENTIFICATION OF THE ZYGOMYCETES

*A.* Microscopic features of Zygomycetes: (*a*) mature sporangium; (*b*) sporangiophore, (*c*) nonseptate hyphae. High power.
*B.* Microscopic appearance of mature sporangia. High power.
*C.* *Rhizopus* sp illustrating characteristic rhizoids. High power.
*D.* *Rhizopus* sp demonstrating nodal placement of sporangiophores arising adjacent to the rhizoids. Low power.
*E.* *Absidia* sp demonstrating internodal placement of the sporangiophores arising between rhizoids. Low power.
*F.* Photomicrograph of the fruiting head of *Syncephalastrum* sp. Note the radial arrangement of cylindrial merosporangia around a central spherical columella. Oil immersion.
*G.* Microscopic features of *Circinella* sp with mature sporangia borne from long, circinoid sporangiophores. High power.
*H.* Photomicrograph of *Cunninghamella* sp illustrating globose columella giving rise to spores arranged in a flowerette.

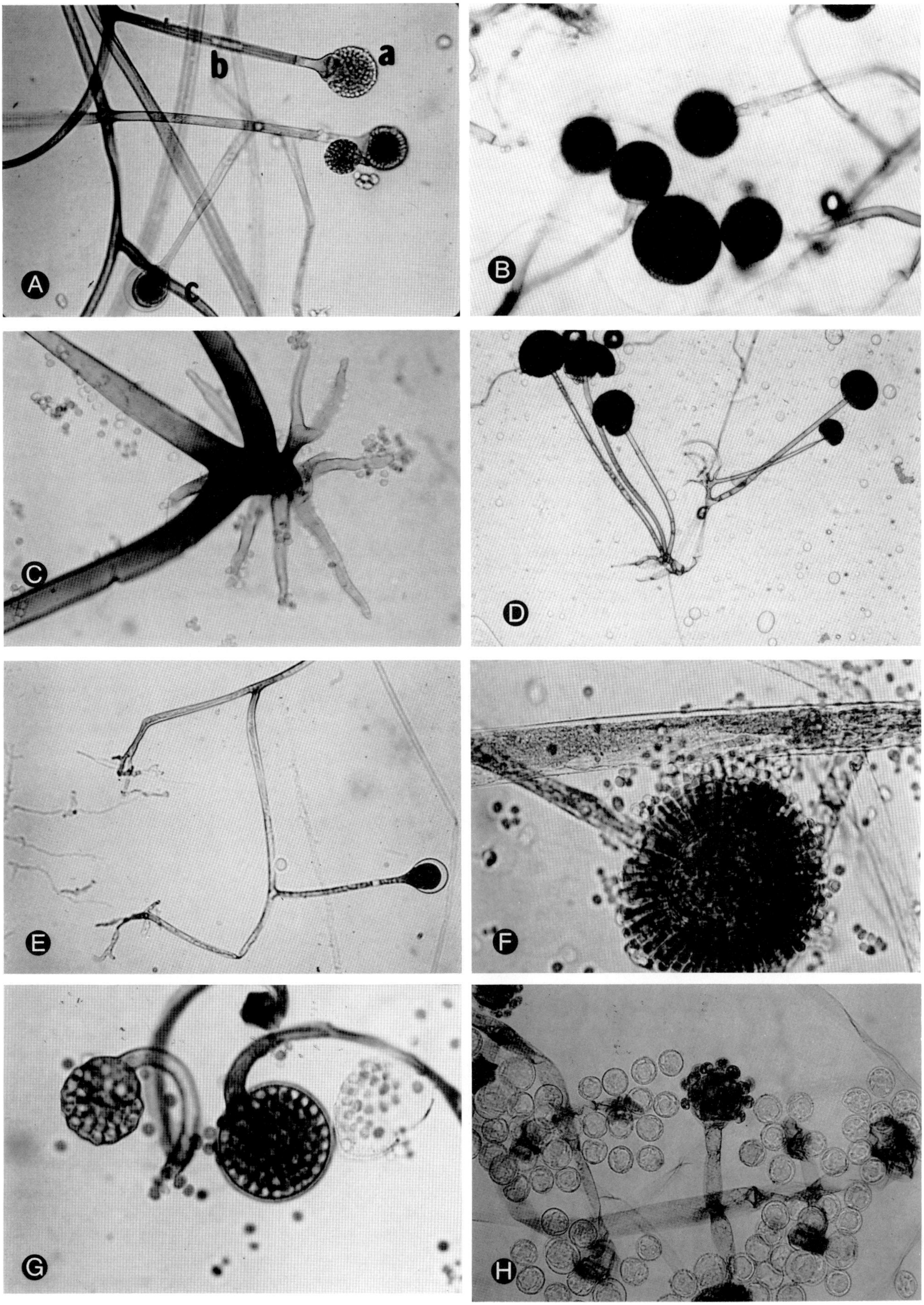
a
b
c
A
B
C
D
E
F
G
H

# PLATE 6.2

## LABORATORY IDENTIFICATION DEMATIACEOUS SAPROBES

*A.* *Alternaria* sp illustrating chains of drumstick-shaped, multicelled macroconidia divided by both transverse and longitudinal septa. High power.

*B.* *Stemphylium* sp showing single arranged, globose, multicelled macroconidia with transecting septa. High power.

*C.* *Epicoccum* sp with characteristic compact aggregates of globose to clavate, multicelled macroconidia divided by transecting septa. High power.

*D.* *Curvularia* sp illustrating, multicelled macroconidia divided by transverse septa with swollen central cells producing a curved appearance. High power.

*E.* Photomicrograph of *Drechslera* sp. Note sympodial arrangement of multicelled macroconidia along a twisted conidiophore. High power.

*F.* *Helminthosporium* sp. Note that the conidia are borne laterally from the sides of a straight conidiophore without sympodial arrangement, characteristics by which this species is differentiated from *Drechslera* sp. High power.

*G.* *Nigrospora* sp illustrating characteristic globose, jet black-staining, single-celled conidia borne atop short, flared conidiophores. High power.

*H.* Photomicrograph of *Aureobasidium* sp illustrating dark-staining, thick-walled arthroconidia and numerous small, elongated blastoconidia. High power.

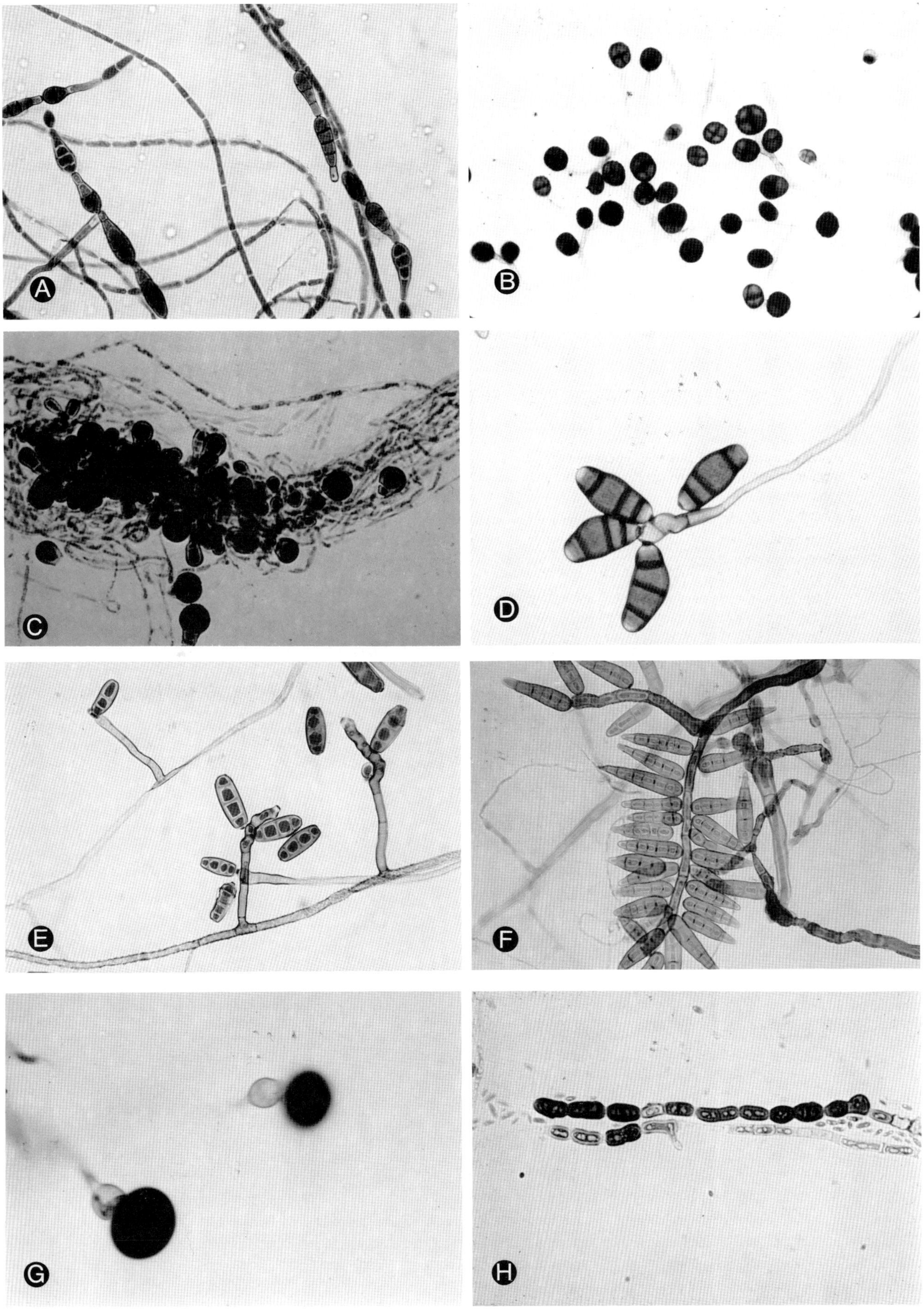
A
B
C
D
E
F
G
H

# PLATE 6.3

## LABORATORY IDENTIFICATION DEMATIACEOUS MOLDS ASSOCIATED WITH SUBCUTANEOUS MYCOSES

*A.* Photomicrograph of *Cladosporium* sp in which oval-shaped conidia are borne in long chains from short, branching conidiophores. Notice the presence of darkened, scar-like areas (disjunctors) on individual conidia at points of attachment. High power.

*B.* *P. verrucosa* illustrating typical flask-shaped phialides producing compact clusters of conidia at their tips. Oil immersion.

*C.* Phialophora-type sporulation. Note collar on tip of phialide within which conidia are being produced internally. High power.

*D.* Characteristic sporulation of *E. jeanselmei*. Note that annellides are not distinguishable; however, the annellophore appears to be tapered at the tip. Oil immersion.

*E.* Darkly pigmented yeast forms of *Exophiala* or *Wangiella* sp. Oil immersion.

*F.* Elongation tube-like phialide of *W. dermatitidis*. Note the absence of a collar at the tip of the phialide. Oil immersion.

*G.* Microscopic appearance of *Fonsecaea pedrosoi* illustrating both the cladosporium and sympodial types of sporulation. High power.

*H.* *F. pedrosoi* illustrating the sympodial (acrotheca) type of sporulation. Oil immersion.

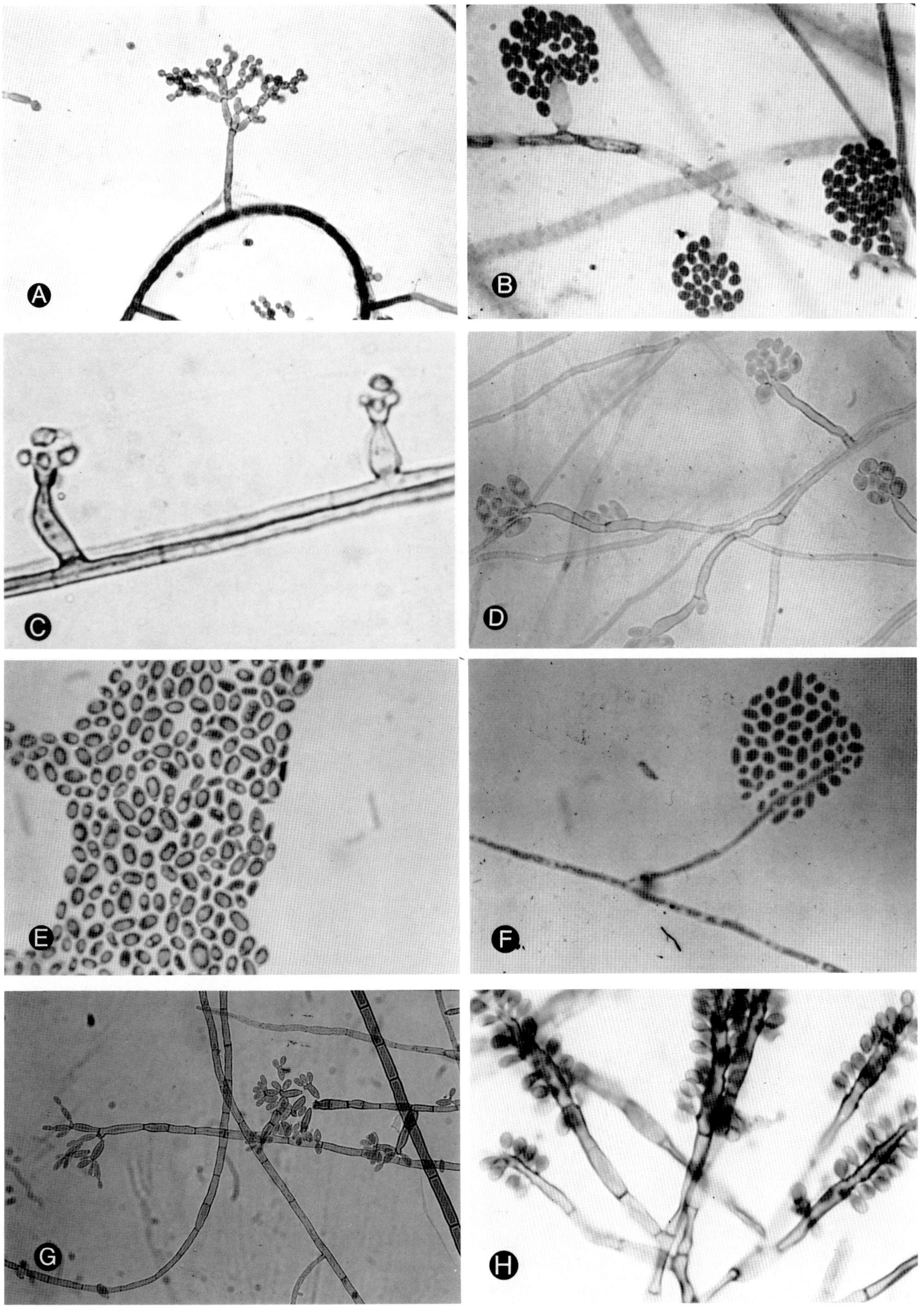
A
B
C
D
E
F
G
H

# PLATE 6.4

## LABORATORY IDENTIFICATION OF *ASPERGILLUS* SPECIES

*A.* Photomicrograph of fruiting head of *A. niger*. Note sporulation of dense clusters of darkly-staining conidia from the entire surface of the vesicle. High power.
*B.* Fruiting head of *A. niger* showing metulae and a single row of phialides with conidia at their tips. High power.
*C.* Fruiting head of *A. fumigatus*. Note clavate vesicle, single row of phialides with sporulation limited to the upper one-half to two-thirds of the vesicle surface. High power.
*D.* Fruiting head of *A. fumigatus* revealing phialides arising from the upper two-thirds of the vesicle surface. High power.
*E.* Fruiting head of *A. flavus* showing metulae and a single row of phialides producing conidia from the entire circumference of the vesicle surface. High power.
*F.* Fruiting heads of *A. terreus* and aleuriospores characteristic of this species. High power.
*G.* Microscopic appearance of a vesicle and a cleistothecium (*arrow*) demonstrating the sexual and the asexual forms of reproduction of certain *Aspergillus* sp. High power.
*H.* Microscopic appearance of spherical Hülle cells produced by certain *Aspergillus* sp. High power.

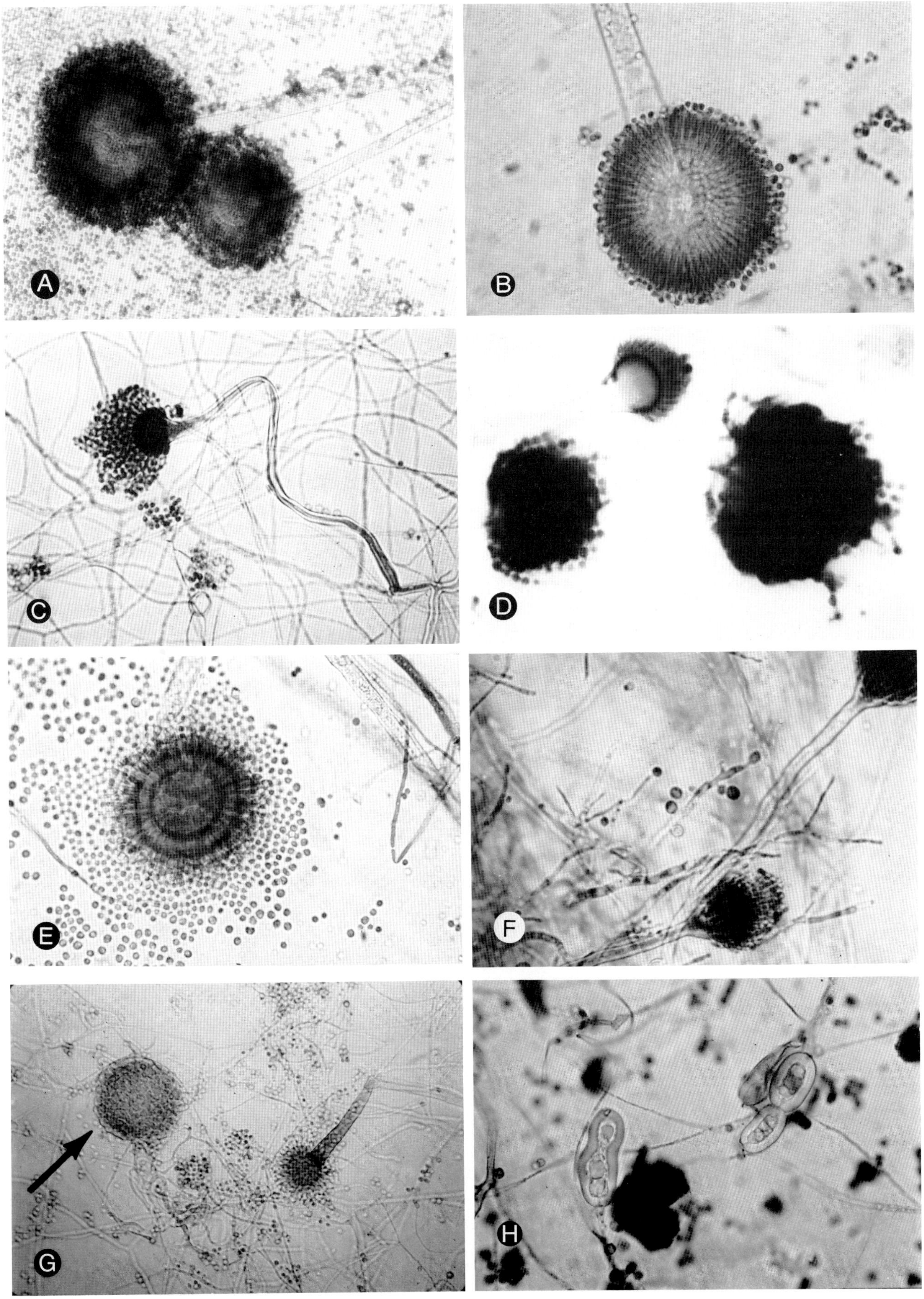
A
B
C
D
E
F
G
H

# PLATE 6.5

## LABORATORY IDENTIFICATION OF HYALINE SAPROBES

*A.* Photomicrograph of fruiting head of *Penicillium* sp. Note branching metulae with phialides that have flattened tips at the points where chains of conidia are derived. High power.
*B.* Enlarged head of *Penicillium* sp showing metulae and a single row of phialides. High power.
*C.* Fruiting head of *Paecilomyces* sp. Note tapered phialides giving rise to chains of conidia. High power.
*D.* Microscopic view of *Scopulariopsis* sp showing long chains of conidia. High power.
*E.* Single head of *Scopulariopsis* sp demonstrating conidia produced at the tips of annellides. High power.
*F.* Fruiting head of *Acremonium* sp illustrating clusters of elliptical conidia at the tips of delicate, almost hair-like phialides. High power.
*G.* Single phialide of *Acremonium* sp producing a compact cluster of conidia at its tip. Oil immersion.
*H.* Fruiting head of *Trichoderma* sp. Note tendency for the conidia to form clusters at the tips of slender, tapered phialides. High power.

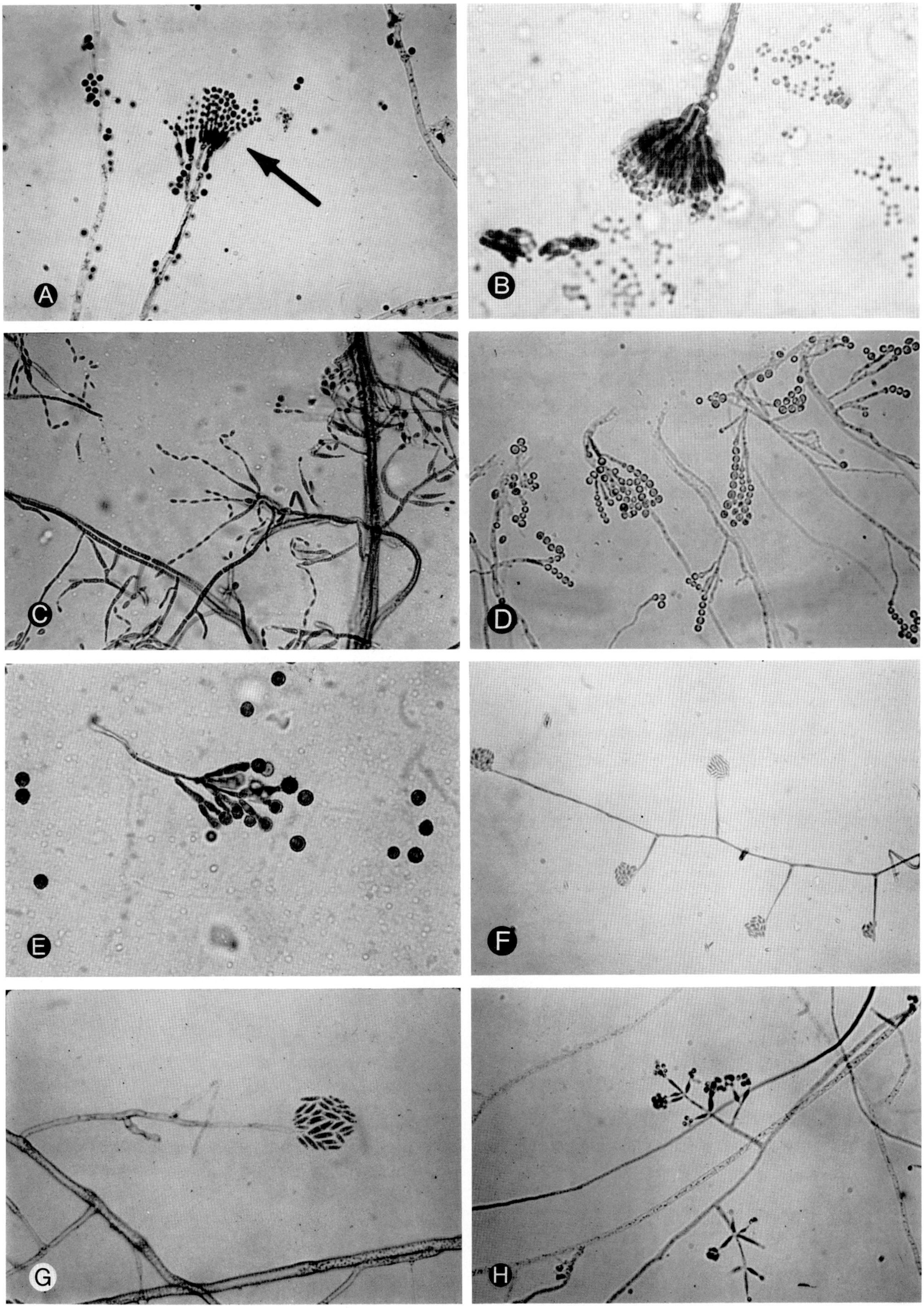
A
B
C
D
E
F
G
H

# PLATE 6.5 (*CONTINUED*)

## LABORATORY IDENTIFICATION OF HYALINE SAPROBES

*I.* Low power view of fruiting head of *Gliocladium* sp. Note clustering of densely packed, spherical conidia borne at the tips of broad, flared phialides. Low power.

*J.* Enlarged view of fruiting head of *Gliocladium* sp with cluster of conidia at the tips of flared phialides. High power.

*K.* Fruiting heads of *Fusarium* sp showing single phialides containing clusters of elliptical conidia at their tips. High power.

*L.* Microscopic view of *Fusarium* sp showing curved, sickle-shaped, multicelled septate conidia. High power.

*M.* Photomicrograph of *Chrysosporium* sp showing single oval to clavate conidia borne from short delicate conidiophores. High power.

*N.* *Sepedonium* sp showing large tuberculate macroconidia characteristic of this genus. High power.

*O.* Photomicrograph of *Scedosporium apiospermum* (*P. boydii*) revealing singly borne, elliptical conidia directly along the hyphae or from the tips of single conidiophores. High power.

*P.* Cleistothecium (sexual sporulation) form of *P. boydii*. Cleistothecia contain asci and ascospores which are not apparent in this photomicrograph (see Plate 5.3*G* and *H*).

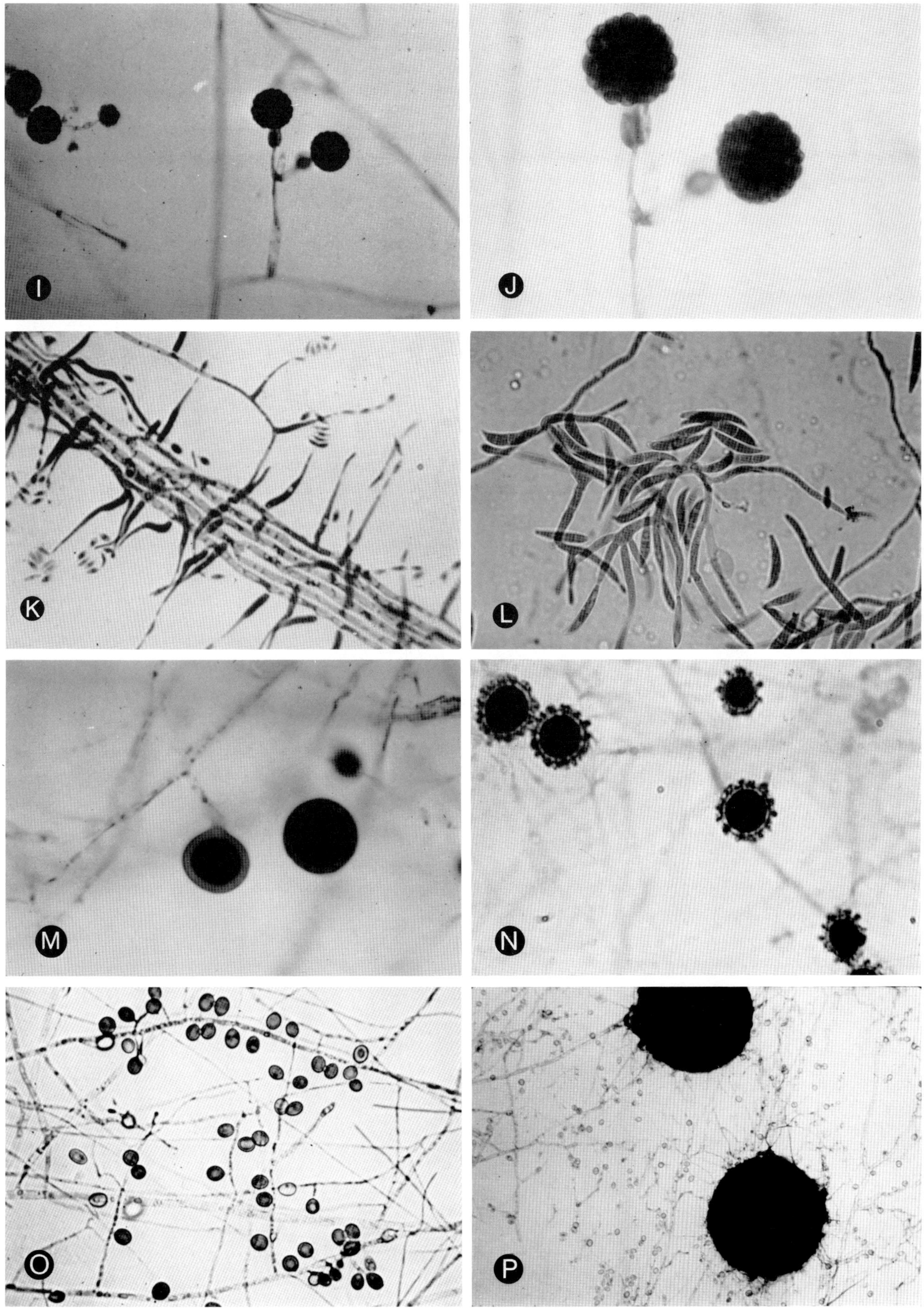
I
J
K
L
M
N
O
P

# PLATE 6.6

## LABORATORY IDENTIFICATION OF DERMATOPHYTIC FUNGI: *Malassezia furfur*, *Epidermophyton*, and *Microsporum*

*A.* Photomicrograph of skin scale containing hyphae and yeast forms characteristic of *Malassezia furfur* (so-called "spaghetti and meatball" effect). Low power.
*B.* Photomicrograph of a PAS-stained skin scale containing hyphae and yeast forms characteristic of *M. furfur*. High power.
*C.* Cluster of smooth-walled, multicelled macroconidia of *E. floccosum*. Microconidia are conspicuous for their absence. Oil immersion.
*D.* Bicelled macroconidia of *M. nanum*, each composed of two cells. Oil immersion.
*E.* Photomicrograph of multicelled, thick and rough-walled macroconidia of *M. canis*. Note spindle shape and curved beak. Oil immersion.
*F.* Broad macroconidia of *M. gypseum* illustrating rounded tips, the chief differential feature from the more pointed macroconidia of *M. canis*. Oil immersion.
*G.* Distorted macroconidium of *M. audouinii*, a rare finding in cultures of this organism. High power.
*H.* Photomicrograph of *M. audouinii* demonstrating lack of conidial formation. Note presence of a terminal chlamydospore, often the only structures produced by this organism. High power.

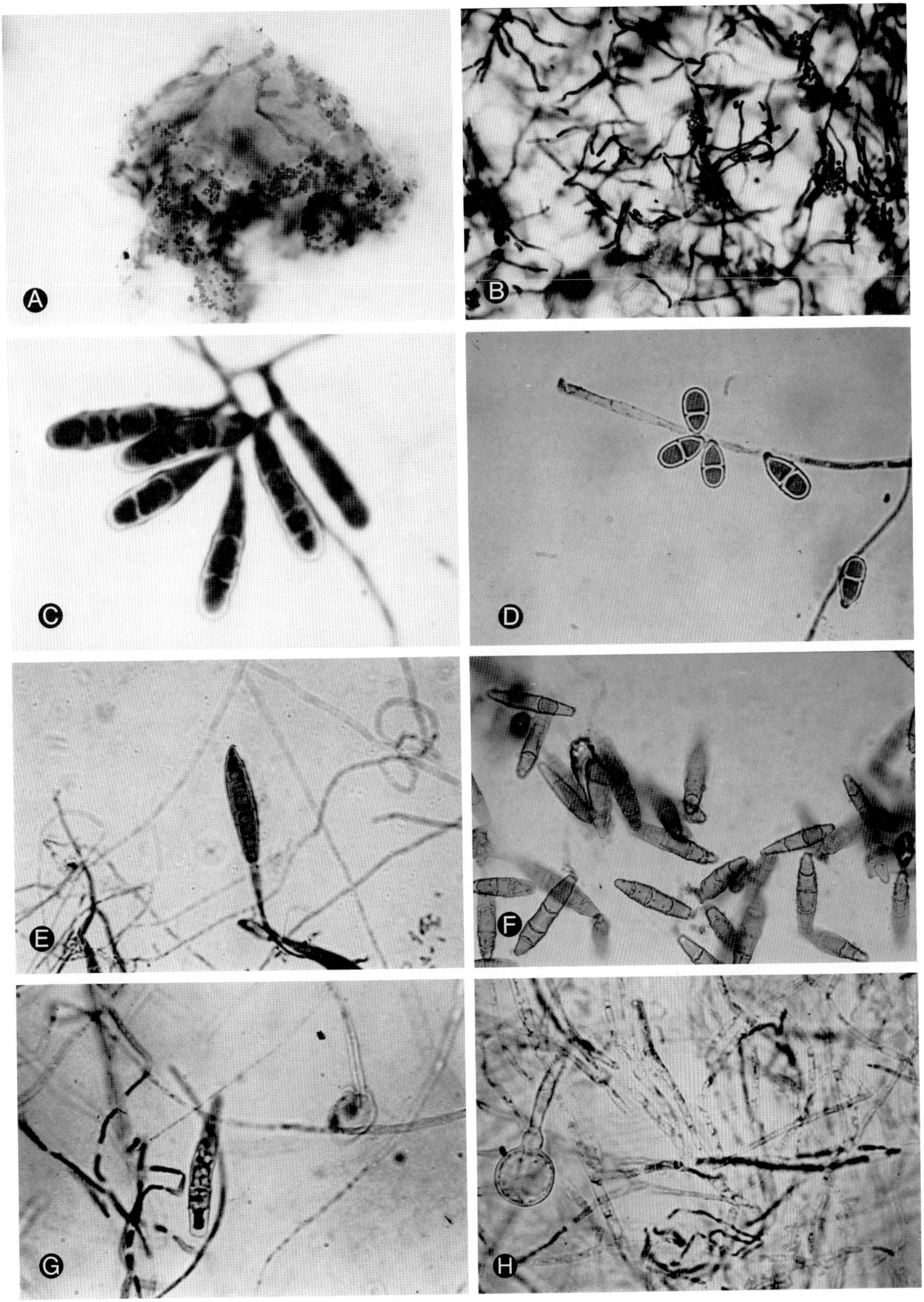
A
B
C
D
E
F
G
H

# PLATE 6.7

## LABORATORY IDENTIFICATION OF DERMATOPHYTIC FUNGI: *Trichophyton* species

*A.* Photomicrograph of *T. mentagrophytes* illustrating loosely held clusters of spherical microconidia and spiral hyphae. High power.
*B.* Microscopic view of *T. rubrum* showing pyriform microconidia borne singly in a sleeve-like arrangement from opposite sides of the hyphae. High power.
*C.* Pencil-shaped, thin and smooth-walled macroconidia characteristic of the genus *Trichophyton*. High power.
*D.* Oil immersion photomicrograph of a lactophenol-cotton blue mount of a hair shaft infected with *T. mentagrophytes* obtained from a hair baiting test. Note wedge-shaped areas of penetration of the hair shaft at points of infection. *T. rubrum* is incapable of invading the hair shaft. Oil immersion.
*E.* Microscopic view of *T. tonsurans* illustrating variation in size of microconidia. Note presence of swollen "ballooned" microconidia admixed with those that are clavate or tear-shaped. High power.
*F.* *T. verrucosum* revealing formation of microconidia along the sides of hyphae. Formation of macroconidia with this organism is very uncommon. High power.
*G.* Characteristic antler hyphae of *T. schoenleinii*. Formation of conidia is virtually never seen with this organism. High power.
*H.* *T. violaceum* showing tortuous hypha containing chlamydospores aligned in chains. High power.

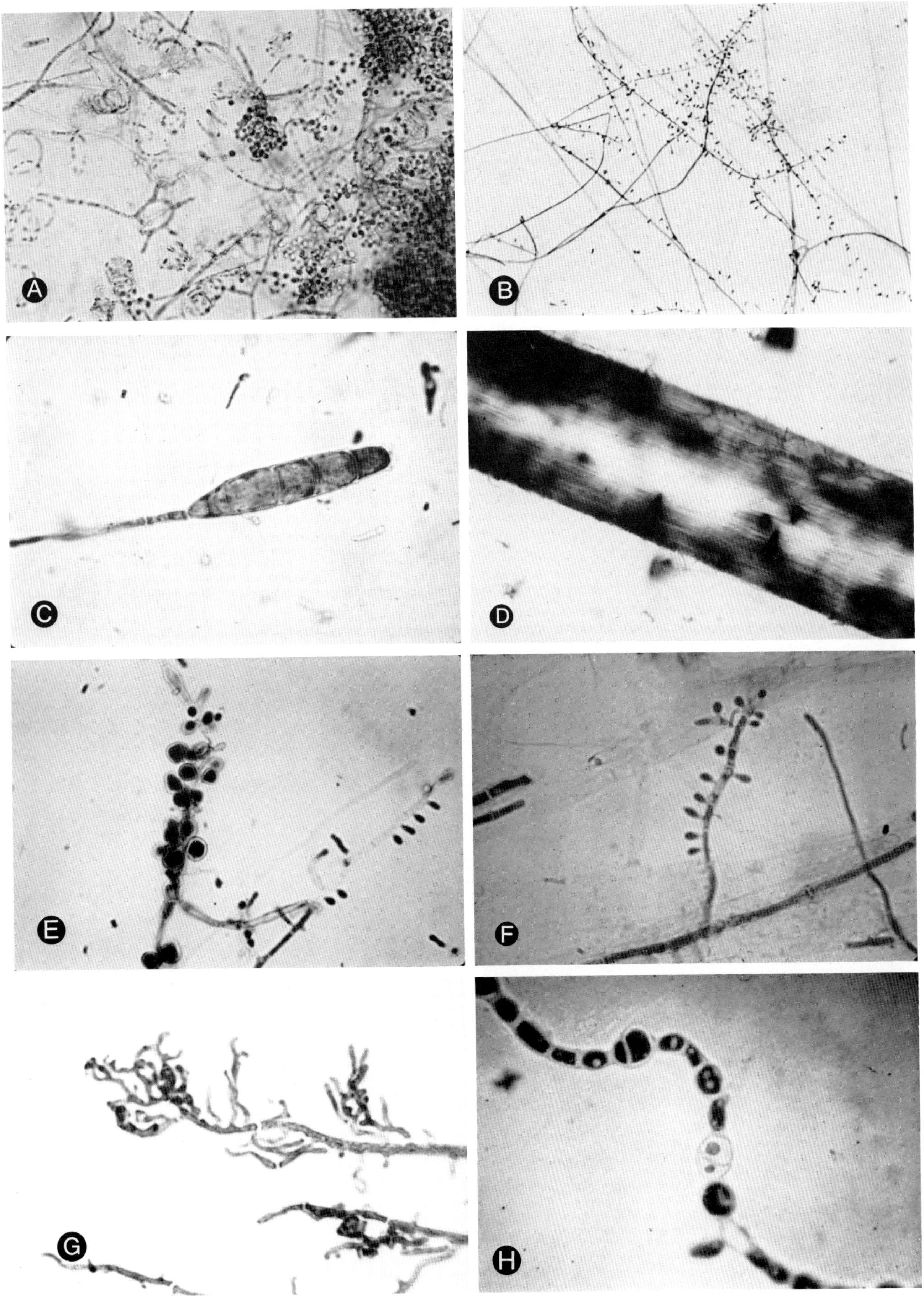
A
B
C
D
E
F
G
H

# PLATE 6.8

## LABORATORY IDENTIFICATION OF DIMORPHIC MOLDS: *Blastomyces dermatitidis* and *Paracoccidioides brasiliensis*

*A.* Photomicrograph of mold form of *B. dermatitidis* showing small, branching hyphae. Delicate hyphae with this appearance are commonly seen with all of the dimorphic fungi. High power.

*B.* Microscopic view of mold form of *B. dermatitidis* showing single clavate conidia borne singly from separate conidiophores, producing a so-called "lollypop" effect. Note parallel arrangement of hyphae, an arrangement commonly seen in cultures grown on medium containing blood. High power.

*C.* Mold form of *B. dermatitidis* showing numerous conidia from single conidiophores characteristic of this fungus. High power.

*D.* Detailed photomicrograph showing pyriform conidia from the tips of short, single conidiophores characteristic of *B. dermatitidis*. High power.

*E.* Yeast form of *B. dermatitidis* showing yeast cells with double-contoured walls and single bud attached to mother cell by a broad base. High power.

*F.* Photomicrograph of *B. dermatitidis* showing an intermediate stage of conversion from the mold to the yeast form. Note resemblance of swollen hyphae to the yeast forms. It should be noted that the yeast conversion stage of *H. capsulatum* can show microscopic appearance similar to that illustrated here. High power.

*G.* Mold form of *P. brasiliensis* revealing delicate hyphae and single conidia, a microscopic appearance similar to that seen with the mold form of *B. dermatitidis*. High power.

*H.* Yeast form of *P. brasiliensis* showing a large yeast cell surrounded by multiple buds. In some instances the parent yeast cell will be totally surrounded by smaller budding yeast cells. High power.

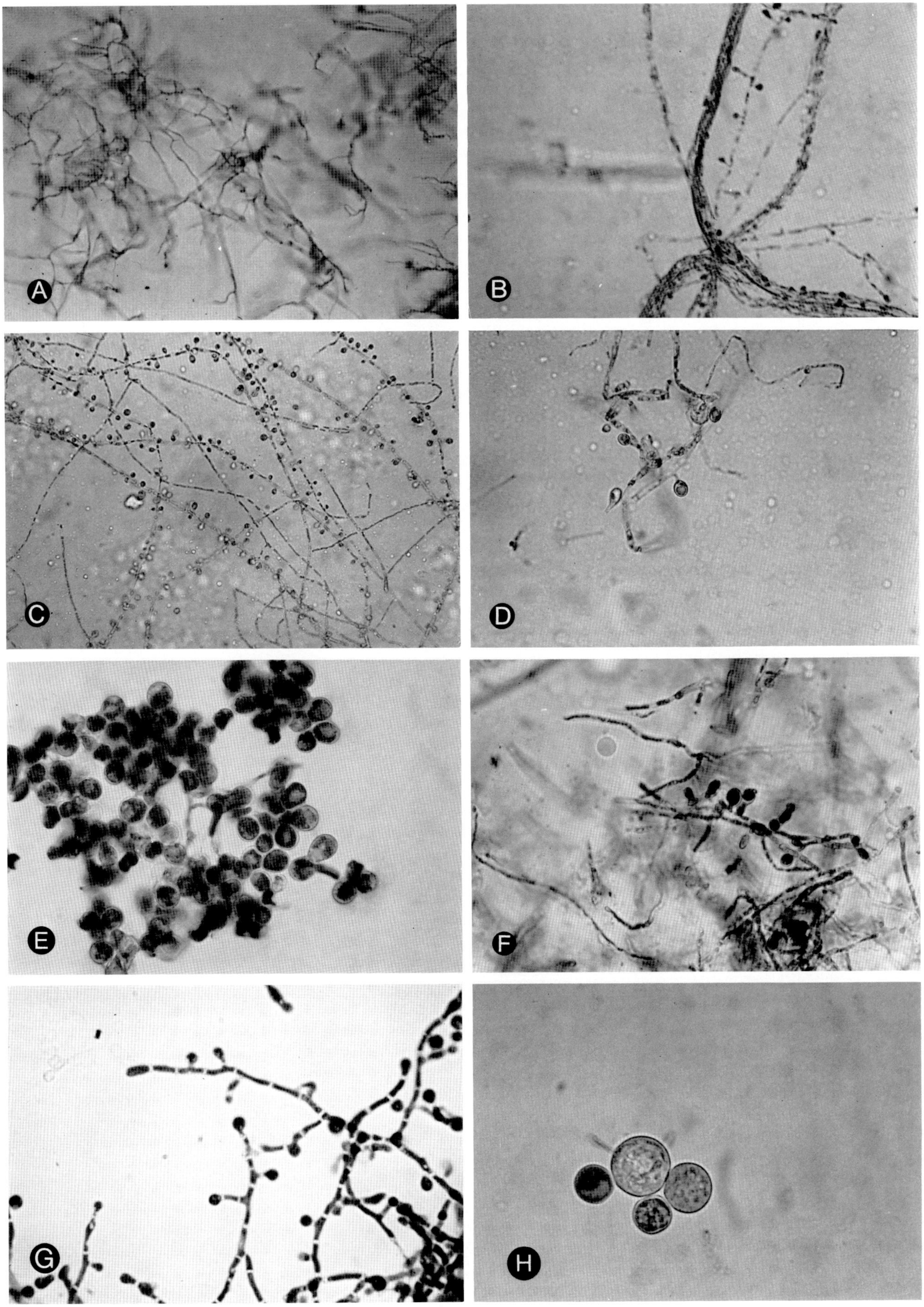
A
B
C
D
E
F
G
H

# PLATE 6.9

## LABORATORY IDENTIFICATION OF DIMORPHIC MOLDS: *Histoplasma capsulatum*

*A.* Photomicrograph of mold form of *H. capsulatum* showing delicate branching hyphae, a microscopic appearance characteristic of all dimorphic fungi. In many instances, the hyphae may be clustered in a rope-like fashion. High power.

*B.* Mold form of *H. capsulatum* showing production of microconidia. Note that the microconidia are similar to those of *B. dermatitidis* and can also resemble the microconidia of dermatophytes, particularly *T. rubrum*. High power.

*C.* High power microscopic view of the mold form of *H. capsulatum* illustrating the microconidia characteristic of this species. High power.

*D.* Mold form of *H. capsulatum* showing immature macroconidia which have a smooth wall at this stage of development. In most instances smooth macroconidia will become tuberculate after an additional period of incubation. High power.

*E.* Mold form of *H. capsulatum* exhibiting formation of both microconidia and macroconidia. High power.

*F.* Mold form of *H. capsulatum* illustrating tuberculate macroconidia characteristic of this species. High power, phase contrast.

*G.* Mold form of *H. capsulatum* illustrating macroconidia arranged in chains, a picture often seen with primary isolates. High power.

*H.* Photomicrograph of mount prepared from a yeast colony of *H. capsulatum* illustrating small yeast cells ranging from 2–5 $\mu$m in diameter. Culture presented here was obtained from the slant of a brain-heart infusion agar containing blood incubated at 35° C. Oil immersion.

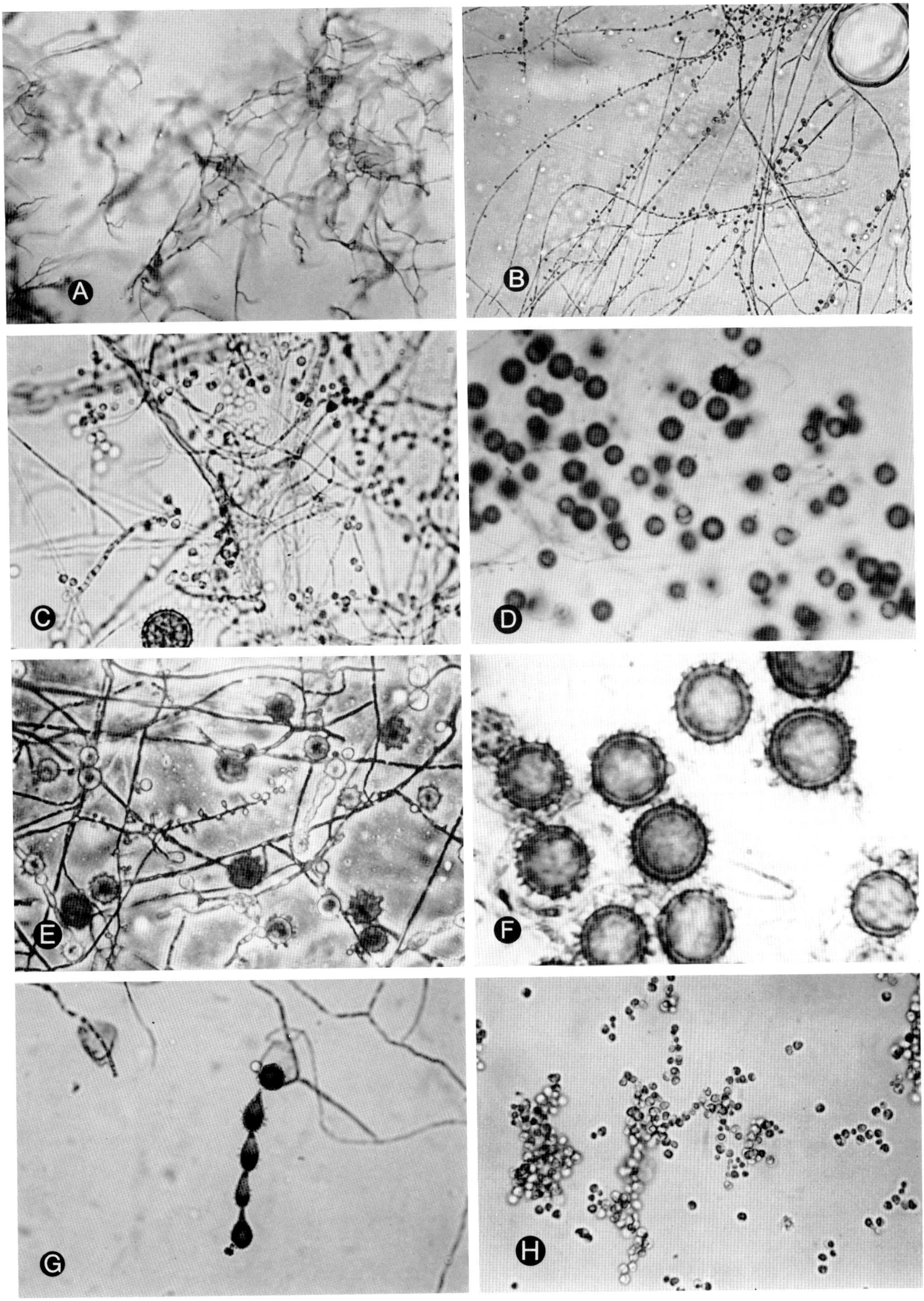
A
B
C
D
E
F
G
H

# PLATE 6.10

## LABORATORY IDENTIFICATION OF DIMORPHIC MOLDS: *Coccidioides immitis* and *Sporothrix schenckii*

*A.* Photomicrograph of mold form of *C. immitis* showing racquet hyphae formation, a picture often seen in young cultures along with right angle branching. High power.
*B.* Mold form of *C. immitis.* Note barrel-shaped arthroconidia separated by empty cells, giving an alternate staining effect. High power.
*C.* Microscopic view of a germinating spherule of *C. immitis* after incubation in a moist chamber after 24 hours. Note production of multiple hyphal forms penetrating through the spherule wall. High power.
*D.* Microscopic view of *Gymnoascus* sp illustrating alternately staining arthroconidia that tend to be more rectangular than those of *C. immitis*. This is an unreliable microscopic feature, however, and cannot be used as the sole means for differentiating between these two fungi. High power.
*E.* Arthroconidia as may be seen with either *Geotrichum* sp or *Trichosporon* sp illustrating rectangular, evenly staining arthroconidia. High power.
*F.* Characteristic flowerette arrangement of small, pyriform microconidia of the mold form of *S. schenckii*. The characteristic thread-like attachments of the conidia to the conidiophore or to the hyphae are difficult to see at this magnification. High power.
*G.* *S. schenckii* mold form illustrating both the flowerette and the sleeve-like arrangement of microconidia along the sides of the hyphae. Note that conidia arranged in sleeve-like fashion appear to be darkly pigmented.
*H.* Microscopic view of the yeast cells of *S. schenckii*. Note that the yeast cells may be oval or elongated (characteristically cigar-shaped) and may exhibit multiple buds. High power.

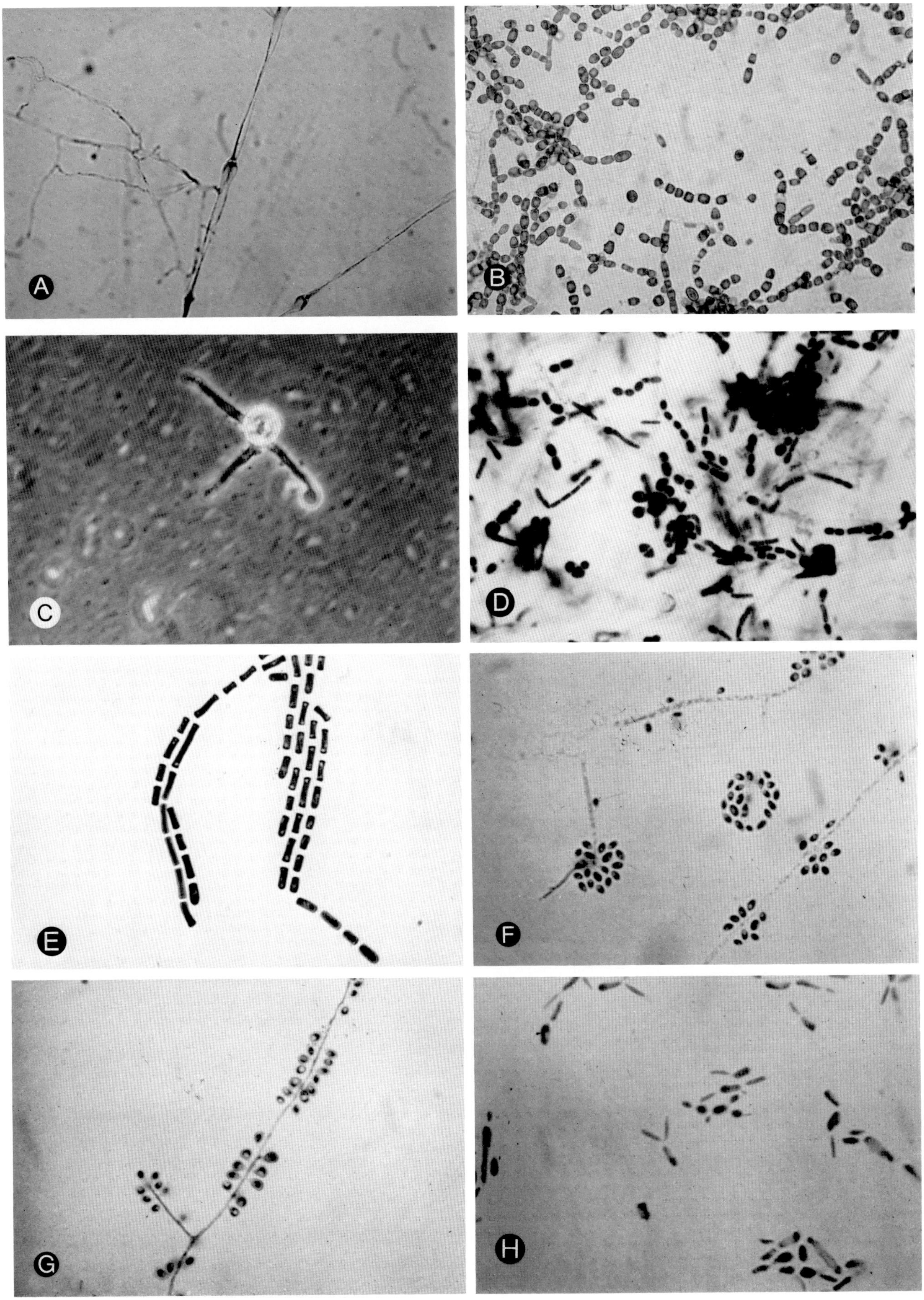
A
B
C
D
E
F
G
H

# Chapter 7

# LABORATORY IDENTIFICATION OF YEASTS AND YEAST-LIKE ORGANISMS

In the past few years, a significant increase in the number of yeast infections has been reported in the medical literature. Most infections have been attributed to ***Cryptococcus neoformans*** and ***Candida albicans***; however, many other species have also been implicated. Yeast infections are seen primarily in patients who are debilitated or have chronic disease or who are immunocompromised with host defense mechanisms altered by the administration of corticosteroids, cytotoxic and immunosuppressive agents, or the long-term or indiscriminate use of antibiotics (42). It is now recognized that patients having gastrointestinal surgery and patients on long-term intravenous therapy without adequate catheter care may also acquire infections caused by yeast and yeast-like organisms. Microbiologists must be aware that any species of yeast is potentially pathogenic in these patients.

## WHEN TO IDENTIFY YEASTS

Is it necessary to identify yeasts recovered from all clinical specimens? This question is commonly asked by microbiologists in clinical laboratories where up to 50% or more of all specimens submitted for fungal culture contain yeasts. Respiratory secretions are the most common specimens from which yeasts are recovered. It has been well demonstrated that the presence of yeasts, excluding *C. neoformans*, usually represent normal endogenous flora and a complete identification on a routine basis is unnecessary (62).

The repeated recovery of different species of yeasts from multiple specimens from the same patient almost certainly indicates colonization and infrequently may indicate local or systemic infection. In this situation, the physician should be consulted on the next steps to be taken since the complete identification of the etiological agent(s) may be important in making a therapeutic decision. If a physician feels that an immunocompromised patient has evidence of a primary pneumonia produced by yeast or yeast-like organisms, a complete identification should be attempted. Other important situations that justify the complete identification of yeasts include their recovery from normally sterile fluids such as blood, cerebrospinal, synovial, abdominal or thoracic. If yeasts are recovered simultaneously from more than one body site, each isolate should be completely identified. If all isolates are of the same species, disseminated infection can be suspected.

In view of the aforementioned information, each laboratory director must decide when to spend the time and expense to perform complete identification procedures on yeasts recovered from clinical specimens. For these specimens where yeast isolates are considered to be normal flora, a simple screen for urease production will detect the presence of *C. neoformans*. If the test is negative, a report stating "yeast present, not *Cryptococcus neoformans*" can be issued and further studies can be bypassed unless a special request is received from the physician.

## PRACTICAL APPROACH TO THE LABORATORY IDENTIFICATION OF YEASTS

Since the publication of the second edition of this text, the laboratory approach to the identification of yeasts and yeast-like organisms recovered from clinical specimens has shifted from conventional tests to the use of commercially available systems. The development of the latter has provided laboratories of all sizes with the capability of using standardized methods. Regardless of whether a commercially available system or conventional tests are used, the germ tube test is helpful to screen for the presence of *Candida albicans* (see page 149). If the test is positive, the identification of *C.*

*albicans* may be made and further testing is not required. The germ tube test thus allows the laboratory to identify rapidly approximately 75% of all the yeasts recovered from clinical specimens.

The rapid urease test is useful for screening for urease-producing yeasts recovered from the respiratory tract and from other specimens (73). Alternately a tube of Christensen's urea agar can be heavily inoculated with the test organism. If the urease test is positive, other rapid conventional methods or commercially available systems may be used to identify *C. neoformans* definitively.

### Commercial Systems

Three systems, the **API 20C,** the **API-Yeast-Ident** (Analytab Products, Inc, Plainview, NY) and the **Uni-Yeast-Tek System** (Flow Laboratories, Roslyn, NY) will be described. In addition, the **Minitek System** (BioQuest, BBL, Cockeysville, MD), primarily designed for the identification of the *Enterobacteriaceae*, may be employed for the identification of yeasts as well.

Two automated systems, the **AMS** (Vitek Systems, Hazelwood, MO) that utilizes a specific biochemical card and the **Quantum** (Abbott Laboratories, Diagnostic Division, Dallas, TX) that uses a yeast cartridge, are currently being evaluated in many laboratories. These automated systems employ a series of separate plastic wells, each of which contains a dehydrated carbohydrate or other biochemical substrate. After inoculation with a suspension of the yeast to be identified, the inoculated cards are placed in the combination incubation and reading module of the respective instrument and light scatter diodes read at frequent intervals for the presence of turbidity. The electronic signals are fed into a microprocessor from which computer-generated organism codes are printed out on a card. The systems are relatively expensive and are limited to use in laboratories with a sufficiently heavy work load to make their use cost effective. These systems will not be discussed further here and interested readers are requested to contact the manufacturers for specific details.

#### API 20C STRIP

The system is similar in design to the API 20E strip, widely used for the identification of the *Enterobacteriaceae* and nonfermentative gram-negative bacilli. The strip includes 20 microcupules, 19 of which contain dehydrated carbohydrate substrates for performing utilization studies (Fig. 1 of Color Plate 6). The substrates contained within the strip include:

Dextrose
Glycerol
2-keto-d-gluconate
L-arabinose
Xylose
Adonitol
Xylitol
Galactose
Sorbitol
Methyl-d-glucoside
N-acetyl-d-glucosamine
Cellobiose
Lactose
Maltose
Sucrose
Trehalose
Melezitose
Raffinose

The first cupule in the strip serves as a growth control to ensure that the organism exhibits growth and that the optimal inoculum has been used.

**API 20C Procedure**

- An ampule of the API 20C basal medium is melted by placing it in a boiling water bath followed by cooling to 50°C in a water bath
- A sterile applicator stick is touched gently to the surface of the well-isolated colony from a 24-hour old pure culture of the yeast to be identified and inoculated into the cooled (50°) basal medium; the suspension should be equivalent to a 1+ on the Wickerham card enclosed with the product
- Using a sterile Pasteur pipette, each cupule is completely filled with the yeast suspension, avoiding formation of bubbles
- After inoculation, the strips are placed in an incubation tray containing water to provide a humid atmosphere and the lids are sealed with tape to prevent inadvertant opening
- Strips are incubated at 35°C for 72 hours and are read after 24, 48, and 72 hours of incubation and results are recorded

Reactions are compared to the 0-growth control cupule that serves as a reading standard control for substrate utilization reactions. Cupules showing turbidity significantly heavier than the 0-growth control cupule are considered to be positive. The results of the reactions are converted to a 7-digit biotype profile number and the yeast identification is made from a profile register supplied by the manufacturer. In addition, a computer-assisted identification for those organisms not contained in the profile index is available by calling a toll free number made available upon request by the manufacturer.

Recent studies have shown that the API 20C yeast identification system is satisfactory for the identification of clinically important yeasts (8, 14, 52). It should be noted that the manufacturer's recommendations include the observation of a cornmeal agar plate for determining the characteristic microscopic morphological features for the yeast to be identified, which, in combination with the profile number generated by the test strip, allows one to make a definitive identification of most clinically important yeasts. Table 7.1 is supplied by the manufacturer along with the product.

### UNI-YEAST-TEK SYSTEM

The basic component of this system is a sealed, multicompartment plate containing sterile solid media with each pie-shaped chamber containing one of the following substrates (Fig. 2 of Color Plate 6).

Urea
Nitrate
Nitrate control
Carbohydrate control
Trehalose
Soluble starch
Cellobiose
Raffinose
Maltose
Lactose
Sucrose

The center of the plate contains a well filled with cornmeal agar useful for determining the microscopic morphological features of the yeast to be identified.

**Procedure for the Uni-Yeast-Tek System**

- A small amount of a 24-hour old culture is emulsified in a tube containing 2.5 ml of sterile, distilled water; the suspension may be standardized against a 1+ reaction on a Wickerham card enclosed with the kit

**Table 7.1. BIOCHEMICAL CHARACTERISTICS OF YEASTS TESTED WITH API 20C®[1]**

| ORGANISM | 0[2] | GLU | GLY | 2KG | ARA | XYL | ADO | XLT | GAL | INO | SOR | MDG | NAG | CEL | LAC | MAL | SAC | TRE | MLZ | RAF |
|---|---|---|---|---|---|---|---|---|---|---|---|---|---|---|---|---|---|---|---|---|
| *Candida albicans* | 0 | 100 | 13 | 100 | 2 | 93 | 90 | 97 | 100 | 0 | 98 | 95 | 98 | 0 | 0 | 100 | 99 | 97 | 9 | 0 |
| *Candida ciferrii* | 0 | 100 | 80 | 80 | 100 | 100 | 100 | 60 | 100 | 100 | 60 | 0 | 100 | 40 | 0 | 90 | 100 | 100 | 0 | 100 |
| *Candida guilliermondii* | 0 | 100 | 100 | 96 | 99 | 100 | 98 | 99 | 99 | 0 | 98 | 97 | 100 | 99 | 0 | 96 | 100 | 100 | 94 | 99 |
| *Candida humicola* | 0 | 100 | 100 | 100 | 100 | 100 | 33 | 33 | 100 | 100 | 67 | 100 | 100 | 100 | 100 | 100 | 100 | 100 | 83 | 100 |
| *Candida krusei* | 0 | 100 | 95 | 0 | 0 | 0 | 0 | 0 | 0 | 0 | 0 | 0 | 49 | 0 | 0 | 0 | 0 | 0 | 0 | 0 |
| *Candida lambica* | 0 | 100 | 57 | 0 | 0 | 93 | 0 | 0 | 0 | 0 | 0 | 0 | 96 | 0 | 0 | 0 | 0 | 0 | 0 | 0 |
| *Candida lipolytica* | 0 | 100 | 100 | 0 | 0 | 0 | 0 | 0 | 0 | 0 | 38 | 0 | 79 | 0 | 0 | 0 | 0 | 0 | 0 | 0 |
| *Candida lusitaniae* | 0 | 100 | 95 | 97 | 1 | 64 | 94 | 27 | 25 | 0 | 99 | 97 | 96 | 73 | 0 | 100 | 99 | 100 | 99 | 0 |
| *Candida parapsilosis* | 0 | 100 | 93 | 85 | 98 | 98 | 93 | 4 | 99 | 0 | 99 | 93 | 98 | 0 | 0 | 100 | 100 | 99 | 99 | 0 |
| *Candida paratropicalis* | 0 | 100 | 3 | 100 | 0 | 100 | 100 | 2 | 100 | 0 | 100 | 0 | 100 | 0 | 0 | 51 | 5 | 100 | 0 | 0 |
| *Candida pseudotropicalis* | 0 | 100 | 68 | 0 | 21 | 91 | 3 | 50 | 100 | 0 | 41 | 0 | 0 | 6 | 100 | 0 | 100 | 0 | 0 | 100 |
| *Candida rugosa* | 0 | 100 | 60 | 0 | 3 | 77 | 1 | 37 | 100 | 0 | 97 | 0 | 60 | 0 | 0 | 0 | 0 | 0 | 0 | 0 |
| *Candida stellatoidea* | 0 | 100 | 0 | 100 | 0 | 81 | 8 | 81 | 100 | 0 | 69 | 0 | 100 | 0 | 0 | 73 | 0 | 12 | 0 | 0 |
| *Candida tropicalis* | 0 | 100 | 11 | 100 | 0 | 98 | 99 | 15 | 99 | 0 | 100 | 98 | 98 | 7 | 0 | 100 | 100 | 100 | 99 | 0 |
| *Candida zeylanoides* | 0 | 100 | 100 | 91 | 0 | 0 | 17 | 0 | 0 | 0 | 100 | 0 | 100 | 0 | 0 | 0 | 0 | 65 | 0 | 0 |
| *Cryptococcus albidus var. albidus* | 0 | 100 | 0 | 97 | 94 | 100 | 6 | 6 | 35 | 77 | 85 | 79 | 27 | 100 | 65 | 100 | 100 | 82 | 100 | 44 |
| *Cryptococcus albidus var. diffluens* | 0 | 100 | 0 | 96 | 99 | 100 | 0 | 4 | 0 | 67 | 70 | 41 | 0 | 100 | 0 | 100 | 100 | 96 | 93 | 70 |
| *Cryptococcus laurentii* | 0 | 100 | 15 | 100 | 100 | 100 | 59 | 61 | 100 | 90 | 59 | 71 | 90 | 100 | 98 | 95 | 98 | 95 | 98 | 98 |
| *Cryptococcus neoformans* | 0 | 100 | 0 | 100 | 36 | 96 | 71 | 2 | 97 | 98 | 100 | 97 | 93 | 38 | 0 | 100 | 100 | 76 | 94 | 83 |
| *Cryptococcus terreus* | 0 | 100 | 0 | 100 | 80 | 100 | 0 | 0 | 53 | 40 | 100 | 0 | 93 | 93 | 47 | 7 | 7 | 60 | 0 | 0 |
| *Cryptococcus uniguttulatus* | 0 | 100 | 20 | 100 | 100 | 100 | 20 | 0 | 0 | 99 | 50 | 90 | 90 | 0 | 0 | 100 | 100 | 70 | 100 | 30 |
| *Geotrichum species*[3] | 0 | 100 | 100 | 0 | 0 | 100 | 0 | 0 | 100 | 0 | 80 | 0 | 0 | 0 | 0 | 0 | 0 | 0 | 0 | 0 |
| *Hanseniaspora guilliermondii* | 0 | 100 | 0 | 100 | 0 | 0 | 0 | 0 | 0 | 0 | 0 | 0 | 0 | 100 | 0 | 0 | 0 | 0 | 0 | 0 |
| *Hanseniaspora uvarum* | 0 | 100 | 0 | 100 | 0 | 0 | 0 | 0 | 0 | 0 | 0 | 0 | 0 | 100 | 0 | 0 | 0 | 0 | 0 | 0 |
| *Hanseniaspora valbyensis* | 0 | 100 | 0 | 0 | 0 | 0 | 0 | 0 | 0 | 0 | 0 | 0 | 0 | 100 | 0 | 0 | 0 | 0 | 0 | 0 |
| *Hansenula anomala var. anomala* | 0 | 100 | 100 | 0 | 0 | 47 | 1 | 3 | 3 | 0 | 76 | 100 | 0 | 37 | 0 | 100 | 100 | 97 | 100 | 29 |
| *Kluyveromyces lactis* | 0 | 100 | 100 | 0 | 0 | 0 | 0 | 80 | 100 | 0 | 100 | 100 | 0 | 40 | 100 | 100 | 100 | 100 | 100 | 100 |
| *Prototheca stagnora*[4] | 0 | 100 | 83 | 0 | 0 | 0 | 0 | 0 | 100 | 0 | 0 | 0 | 0 | 0 | 0 | 0 | 17 | 0 | 0 | 0 |
| *Prototheca wickerhamii*[4] | 0 | 100 | 100 | 0 | 0 | 0 | 0 | 0 | 57 | 0 | 0 | 0 | 0 | 0 | 0 | 0 | 0 | 100 | 0 | 0 |
| *Prototheca zopfii*[4] | 0 | 100 | 100 | 0 | 0 | 0 | 0 | 0 | 0 | 0 | 0 | 0 | 0 | 0 | 0 | 0 | 0 | 0 | 0 | 0 |
| *Rhodotorula glutinis* | 0 | 100 | 78 | 67 | 33 | 61 | 50 | 28 | 78 | 0 | 67 | 39 | 0 | 11 | 0 | 100 | 100 | 100 | 100 | 100 |
| *Rhodotorula minuta* | 0 | 100 | 100 | 100 | 97 | 91 | 55 | 3 | 0 | 0 | 61 | 0 | 67 | 36 | 6 | 0 | 91 | 91 | 91 | 0 |
| *Rhodotorula pilimanae* | 0 | 100 | 50 | 1 | 90 | 100 | 90 | 70 | 60 | 0 | 40 | 0 | 0 | 0 | 0 | 0 | 100 | 80 | 0 | 100 |
| *Rhodotorula rubra* | 0 | 100 | 47 | 0 | 74 | 94 | 55 | 41 | 82 | 0 | 27 | 1 | 0 | 1 | 0 | 98 | 100 | 96 | 94 | 98 |
| *Saccharomyces cerevisiae* | 0 | 100 | 22 | 0 | 0 | 0 | 0 | 0 | 93 | 0 | 1 | 39 | 0 | 0 | 0 | 85 | 100 | 56 | 34 | 81 |
| *Sporobolomyces salmonicolor* | 0 | 100 | 8 | 0 | 0 | 0 | 4 | 0 | 4 | 0 | 75 | 0 | 0 | 0 | 0 | 0 | 100 | 100 | 0 | 96 |
| *Torulaspora rosei* | 0 | 100 | 100 | 100 | 0 | 0 | 0 | 0 | 0 | 0 | 85 | 0 | 0 | 0 | 0 | 0 | 100 | 95 | 0 | 85 |
| *Torulopsis candida* | 0 | 100 | 98 | 100 | 70 | 70 | 100 | 90 | 100 | 0 | 100 | 100 | 100 | 100 | 50 | 100 | 100 | 100 | 90 | 90 |
| *Torulopsis glabrata* | 0 | 100 | 15 | 0 | 0 | 0 | 0 | 0 | 0 | 0 | 0 | 0 | 0 | 0 | 0 | 0 | 0 | 97 | 0 | 0 |
| *Trichosporon beigelii (cutaneum)* | 0 | 100 | 49 | 97 | 89 | 100 | 33 | 36 | 93 | 66 | 46 | 89 | 96 | 97 | 96 | 99 | 93 | 89 | 71 | 51 |
| *Trichosporon capitatum* | 0 | 100 | 96 | 0 | 0 | 0 | 0 | 0 | 4 | 0 | 0 | 0 | 1 | 0 | 0 | 0 | 0 | 0 | 0 | 0 |
| *Trichosporon penicillatum* | 0 | 100 | 100 | 0 | 0 | 100 | 0 | 0 | 90 | 0 | 95 | 0 | 0 | 0 | 0 | 0 | 0 | 0 | 0 | 0 |

[1]Figures indicate the percentage of positive reactions after 72 hours of incubation at 30°C.

[2]Negative control; if growth occurs in the control, all assimilations should be compared to the control.

[3]Filamentous fungus whose initial growth may appear yeast-like.

[4]Colorless alga whose growth appears yeast-like.

One drop of the yeast suspension is pipetted into each of the 11 peripheral wells and, using a sterile loop, a small amount of inoculum from an isolated colony is scratched across the surface of the cornmeal agar well in the center of the plate

A sterile coverslip is positioned over the cornmeal agar and the plate is incubated at 30° C for 6 days

Plates are examined daily and the peripheral wells are observed for color changes after 48 and 72 hours

A "logic wheel" is supplied by the manufacturer that assists in the rapid identification; however, in instances where an answer cannot be obtained, a profile index is provided by the manufacturer. Table 7.2, an identification chart supplied by the manufacturer, presents individual substrate reactions. Cooper and associates (14) found this system to be 92% accurate in identifying clinical yeast isolates within 72 hours and 96% accurate when the incubation period was extended to one week. These statistics compare favorably with conventional methods.

Also included with the Uni-Yeast-Tek are three supplementary tubes. One contains a glucose beef-extract broth to promote germ tube formation; another is an agar-based medium containing sucrose for the differentiation of *C. stellatoidea* versus *C. albicans*. The remaining tube (C/N screen) is an agar-based medium containing L-DOPA and is used to detect the phenol oxidase activity of *C. neoformans*.

### API-YEAST-IDENT

The API-Yeast-Ident System is a standardized and sensitive micro method utilizing both miniaturized conventional and chromogenic tests for the identification of yeast and yeast-like organisms. It consists of a series of 20 microcupules containing dehydrated substrates (Fig. 3 of Color Plate 6). The addition of a yeast suspension to each cupule rehydrates the substrate and initiates the reactions. After 4 hours of incubation at 35° C, some reactions are monitored by various indicator systems present within the cupules or by the addition of a specific reagent, cinnamaldehyde.

**Procedure for the API Yeast-Ident System**

A yeast suspension equivalent to a McFarland standard 5 is prepared using a sterile wooden applicator stick and emulsifying the yeast in sterile distilled water (3 ml); yeast suspensions should be used as quickly as possible (within 15 minutes) to inoculate the test strip

Each microcupule is inoculated with 2 or 3 drops of the yeast suspension and strips are placed within the humidity chamber supplied by the manufacturer and are incubated for 4 hours at 35° C in a non-$CO_2$ incubator

After 4 hours of incubation, results of the first 9 tests on the strip are examined for a color change and reactions are compared to the color standards supplied by the manufacturer

One drop of cinnamaldehyde reagent is added to the remaining 11 wells and reactions are recorded at 3 minutes after addition of the reagent

A 7-digit number is generated similar to that used by the API 20C system and identification is made by using profile register supplied by the manufacturer.

Each test strip contains the following substrates:

Urea

P-nitrophenyl phosphate

## Table 7.2. Uni-Yeast-Tek® Chart*

| Reactions for Yeasts | C. albicans | C. famata | C. glabrata | C. guilliermondii | C. humicola | C. intermedia | C. krusei | C. lipolytica | C. lusitaniae | C. parapsilosis | C. pseudo tropicalis | C. rugosa | C. stellatoidea | C. tropicalis |
|---|---|---|---|---|---|---|---|---|---|---|---|---|---|---|
| Germ Tubes | V | – | – | – | – | – | – | – | – | – | – | – | V | – |
| Urea | – | – | – | – | + | – | V | V | – | – | – | – | – | – |
| Sucrose | + | + | – | + | + | + | – | – | + | + | + | – | – | V |
| Lactose | – | V | – | – | V | + | – | – | – | – | + | – | – | – |
| Maltose | + | + | – | + | + | + | – | – | + | + | – | – | + | + |
| Raffinose | – | + | – | + | V | + | – | – | – | – | + | – | – | – |
| Cellobiose | V | + | – | + | V | + | – | – | + | – | V | – | – | V |
| Soluble Starch | + | V | – | V | V | + | – | – | – | – | – | – | + | + |
| Trehalose | V | + | + | + | + | + | – | – | + | + | – | – | + | + |
| Nitrate | – | – | – | – | – | – | – | – | – | – | – | – | – | – |
| Pseudohyphae | + | – | – | V | + | + | + | + | + | + | + | + | + | + |
| Blastospores | + | + | + | + | + | + | + | + | + | + | + | + | + | + |
| Arthrospores | – | – | – | – | – | – | – | – | – | – | – | – | – | – |
| 37° C Growth | + | + | + | + | + | + | + | – | + | + | + | + | + | + |
| 40° C Growth | + | – | + | + | – | – | + | – | + | + | + | + | + | + |
| Glucose F. | + | V | + | + | – | + | + | – | + | + | + | – | + | + |
| Sucrose F. | – | V | – | + | – | + | – | – | V | – | + | – | – | + |
| Inositol | – | – | – | – | + | – | – | – | – | – | – | – | – | – |
| Xylose | + | + | – | + | + | + | – | – | + | + | + | + | + | + |
| Dulcitol | – |  | – | + |  |  | – |  |  | – | – | – | – | – |
| C/N Screen | – | – | – | – | – | – | – | – | – | – | – | – | – | – |
| Chlamydospores | + | – | – | – | – | – | – | – | – | – | – | – | V | – |
| Endospores | – | – | – | – | – | – | – | – | – | – | – | – | – | – |
| Ascospores | – | – | – | – | – | – | – | – | – | – | – | – | – | – |

| Reactions for Yeasts | C. utilis | C. viswanathii | C. zeylanoides | Cr. albidus var. albidus | Cr. albidus var. diffluens | Cr. gastricus | Cr. laurentii | Cr. luteolus | Cr. neoformans | Cr. terreus | Cr. uniguttulatus | D. hansenii | G. candidum | G. capitatum |
|---|---|---|---|---|---|---|---|---|---|---|---|---|---|---|
| Germ Tubes | – | – | – | – | – | – | – | – | – | – | – | – | – | – |
| Urea | – | – | – | + | + | + | + | + | + | + | + | + | – | – |
| Sucrose | + | + | – | + | + | V | + | + | + | – | + | + | – | – |
| Lactose | – | – | – | V | – | V | + | V | – | V | – | V | – | – |
| Maltose | + | + | – | + | + | + | + | + | + | V | + | + | – | – |
| Raffinose | + | – | – | + | + | – | + | + | V | – | V | + | – | – |
| Cellobiose | + | + | V | V | + | + | + | + | V | + | V | + | – | – |
| Soluble Starch | – | + | – | V | V | V | V | – | V | V | V | + | V | – |
| Trehalose | V | + | + | V | + | + | V | + | + | V | V | + | – | – |
| Nitrate | + | – | – | + | + | – | – | – | – | + | – | – | – | – |
| Pseudohyphae | + | + | + | – | – | – | – | – | – | – | – | V | + | + |
| Blastospores | + | + | + | + | + | + | + | + | + | + | + | + | – | + |
| Arthrospores | – | – | – | – | – | – | – | – | – | – | – | – | + | + |
| 37° C Growth | + | + | – | – | + | – | + | – | + | + | – | V | – | + |
| 40° C Growth | + | + | – | – |  | – | – | – | – | – | – |  |  |  |
| Glucose F. | + | + | V | – | – | – | – | – | – | – | – | V | – | – |
| Sucrose F. | + | – | – | – | – | – | – | – | – | – | – | V | – | – |
| Inositol | – | – | – | + | + | + | + | + | + | + | + | – | – | – |
| Xylose | + | + | – | + | + | + | + | + | + | + | + | + | + | – |
| Dulcitol |  |  |  | + | V | – | + | + | + | + | – |  | – | – |
| C/N Screen | – | – | – | – | – | – | – | – | + | – | – | – | – | – |
| Chlamydospores | – | – | – | – | – | – | – | – | – | – | – | – | – | – |
| Endospores | – | – | – |  |  |  |  |  |  |  |  |  |  |  |
| Ascospores | – | – | – | – | – | – | – | – | – | – | – | + | – | – |

| Reactions for Yeasts | G. penicillatum | H. anomala | K. marxianus | P. stagnora | P. wickerhamii | P. zopfii | R. glutinis | R. pilimanae | R. rubra | S. cerevisiae | Sp. salmonicolor | Tr. beigelii | Tr. inkin | Tr. pullulans |
|---|---|---|---|---|---|---|---|---|---|---|---|---|---|---|
| Germ Tubes | – | – | – | – | – | – | – | – | – | – | – | – | – | – |
| Urea | – | V | V | – | – | – | + | V | + | – | + | V | – | + |
| Sucrose | – | + | + | V | – | – | + | + | + | + | + | V | + | + |
| Lactose | – | – | V | – | – | – | – | – | – | – | – | + | + | V |
| Maltose | – | + | – | – | – | – | V | – | + | V | – | V | + | + |
| Raffinose | – | + | + | – | – | – | V | + | + | V | V | V | – | + |
| Cellobiose | – | + | + | – | – | – | V | V | V | – | – | + | + | V |
| Soluble Starch | – | + | – | V | – | V | – | – | – | – | – | V | + | V |
| Trehalose | – | + | – | – | + | – | V | + | V | V | + | V | + | + |
| Nitrate | – | + | – | – | – | – | + | – | – | – | + | – | – | + |
| Pseudohyphae | + | V | + | – | – | – | V | – | V | V | + | + | + | + |
| Blastospores | + | + | + | + | + | + | + | + | + | + | + | + | + | + |
| Arthrospores | + | – | – | – | – | – | – | – | – | – | – | + | + | + |
| 37° C Growth | + | V | + | – | + | + | + | + | + | V | – | + | + | – |
| 40° C Growth |  |  | – |  |  |  | – | – | – |  |  | + | + | – |
| Glucose F. | V | + | + |  |  |  | – | – | – | + | – | – | – | – |
| Sucrose F. | – | + | + | – |  |  | – | – | – | + | – | – | – | – |
| Inositol | – | – | – | – | – | – | – | – | – | – | – | V | + | V |
| Xylose | + | V | + | – | – | – | V | + | + | – | V | + | + | V |
| Dulcitol | – |  |  |  |  |  | – |  | – | – |  | + | – | – |
| C/N Screen | – | – | – | – | – | – | – | – | – | – | – | – | – | – |
| Chlamydospores | – | – | – | – | – | – | – | – | – | – | – | – | – | – |
| Endospores |  |  |  | + | + | + |  |  |  |  |  |  |  |  |
| Ascospores | – | + | + | – | – | – | – | – | – | + | – | – | – | – |

KEY + >90%
– >90%
V variable

* Reproduced with permission of Flow Laboratories, Inc., McLean, VA. A revised percentage chart and codebook will be available in 1985.

P-nitrophenyl-β-D-fucoside
P-nitrophenyl-β-D-glucoside
P-nitrophenyl-β-D-galactosaminide
P-nitrophenyl-α-D-glucoside
O-nitrophenyl-β-D-xyloside
Proline-p-nitroanilide
Indoxyl acetate
Glycine-β-naphthylamide
Proline-β-naphthylamide
Tryptophan-β-naphthylamide
Hydroxyproline-β-naphthylamide
Isoleucine-β-naphthylamide
Valine-β-naphthylamide
Leucyl-glycine-β-naphthylamide
Histidine-β-naphthylamide
Cystine-β-naphthylamide
Tyrosine-β-naphthylamide
Glycl-glycyl-β-naphthylamide

In some instances it will be necessary to supplement the results of the test strip with additional conventional biochemical tests and an assessment of the microscopic morphological features found on cornmeal agar. The organism identifications using this strip will be computer assisted based on a large data base provided by the manufacturer. This system will provide for the rapid identification of yeasts encountered in the clinical laboratory.

## GERM TUBE TEST

As previously mentioned, greater than 75% of all yeasts encountered in the clinical laboratory are *C. albicans.* The germ tube test provides an identification within 3 hours.

### Procedure for the Germ Tube Test

A very small inoculum from an isolated yeast colony is suspended in 0.5 ml of sheep serum (rabbit plasma used for prothrombin testing and normal human serum are satisfactory alternatives). It is important that the inoculum be quite small to avoid false-negative results.

The inoculated tubes are incubated at 35° C for 3 hours.

After incubation, a drop of the yeast suspension is placed on a clean microscope slide, covered with a coverslip and examined under low power magnification for the presence of germ tubes. Germ tubes are appendages ½ the width and 3–4 times the length of the yeast cells from which they arise (Fig. 7.1).

Most germ tubes appear as hyphal-like extensions from the yeast cell (Fig. 7.1) without a constriction at a point of origin from the cell. However, it is not uncommon to see germ tubes that show constrictions and their presence or absence is of little significance in making an identification. *C. tropicalis* should be used as a negative control; however, this species can form pseudogerm tubes after 3 hours of incubation, a reminder why this time span should not be exceeded when interpreting the test.

Many microbiologists believe that *C. stellatoidea* is a variant of *C. albicans* and usually make no distinction between the two. The former may be distinguished from the latter by its inability to utilize sucrose. In this text, we consider *C. albicans* and *C. stellatoidea* to be the same organism.

## RAPID UREASE TEST

The rapid urease test (73) is useful for screening for the presence of cryptococci in clinical specimens. Difco Urea-R broth is recommended. Each vial is reconstituted with 3 ml of sterile, distilled water on the day the test is to be performed and at the time of testing 3 or 4

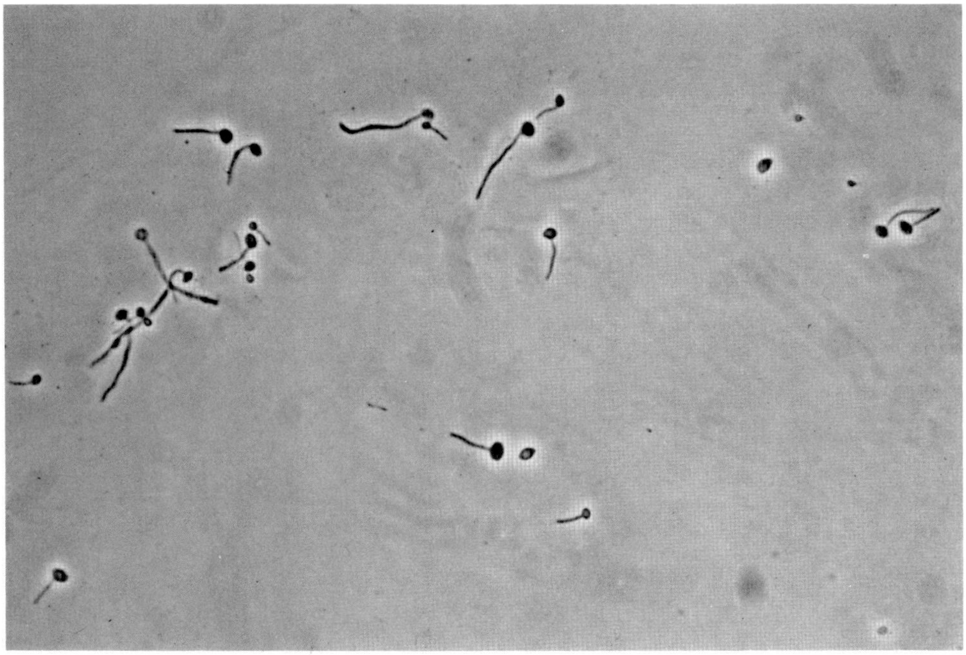

**Figure 7.1.** Microscopic view of a germ tube test illustrating formation of germ tubes characteristic of *C. albicans.*

drops of this suspension are placed into wells of a microtube plate.

**Procedure for the Rapid Urease Test**

A heavy inoculum of the yeast colony to be tested (excluding pink yeasts) is transferred to the microtube well containing the Urea-R broth. A positive control (*C. neoformans*) and a negative control (*C. albicans*) are also tested in parallel.

The inoculated wells of the microtube plate are covered with Scotch tape and the plate is incubated for 4 hours at 37° C.

After incubation the wells are observed for the development of a pink color.

Alternatively, a large inoculum from a single colony of the yeast to be identified is placed on the surface of the upper portion of a slant of Christensen's urea agar. The tube is incubated at 37° C. Urease-producing yeasts will produce a detectable color change within several hours. Cultures should be held for 72 hours before reporting as negative. Those yeasts producing urease can be further identified by setting up other tests including carbohydrate assimilation studies.

## C/N SCREEN

C/N-Screen (Flow Laboratories, Roslyn, NY) is a rapid test useful for the presumptive identification of *C. neoformans* (15) (Fig. 4 of Color Plate 6). This system uses L-DOPA as substrate and is designed to detect phenoloxidase production by the yeast being tested. The development of a brown pigmentation (melanin) by the organism being tested indicates phenoloxidase activity. This test has been found to perform with a sensitivity of 92% and a specificity of 99% in providing a presumptive identification of clinical isolates of *C. neoformans* (15).

**Procedure for the C/N Screen Test**

Inoculate a C/N-Screen tube with a loopful of the yeast to be identified onto the surface of the slant.

Incubate tubes in an upright position at 25° C for 24–48 hours.

Observe for the presence of black growth on the slant (Fig. 4 of Color Plate 6).

Other tests that may be used in combination with these previously mentioned will be described later in this chapter (rapid selective urease test [92], rapid nitrate reductase test [36] and L-DOPA-ferric citrate test [39]). Specific uses for these tests will also be described.

## CONVENTIONAL TESTS USEFUL FOR THE IDENTIFICATION OF YEASTS AND YEAST-LIKE ORGANISMS

Despite the availability of commercial yeast systems, some clinical microbiology laboratories prefer to use conventional laboratory tests. Figure 7.2 is a flow sheet depicting a practical approach to the identification of yeasts using conventional methods. As previously mentioned, the germ tube test is useful to identify *C. albicans;* however, several additional procedures are needed to identify other species.

## CORNMEAL AGAR MORPHOLOGY

Traditionally, microbiologists have used cornmeal agar for the detection of chlamydospores produced by *C. albicans.* Currently, the advocation of cornmeal agar morphology has been expanded to include a presumptive identification of commonly encountered *Candida* sp based on microscopic morphological features and the differentiation of the genera *Cryptococcus, Saccharomyces, Geotrichum,* and *Trichosporon.* The observation of the microscopic morphological features of yeasts on cornmeal agar is necessary not only for conventional methods but for the identification using the API 20C System or the Uni-Yeast-Tek System.

The formation of characteristic microscopic features of yeasts and yeast-like organisms growing on cornmeal agar is enhanced by the addition of Tween-80 (polysorbate 80) which reduces the surface tension of the medium and promotes optimal production of hyphae and blastoconidia. Trypan blue is included in the medium to aid visual observations.

**Cornmeal Agar Procedure**

A plate of cornmeal Tween-80 agar is inoculated by making three parallel cuts ½ inch apart into the medium at a 45° angle.

Replace the lid and incubate the plate for 24–48 hours at 30° C.

After incubation the plates are removed and appropriate areas of growth are examined under low and high power objectives of the microscope. A thin coverslip may be positioned on the agar surface in the area to be examined to prevent the

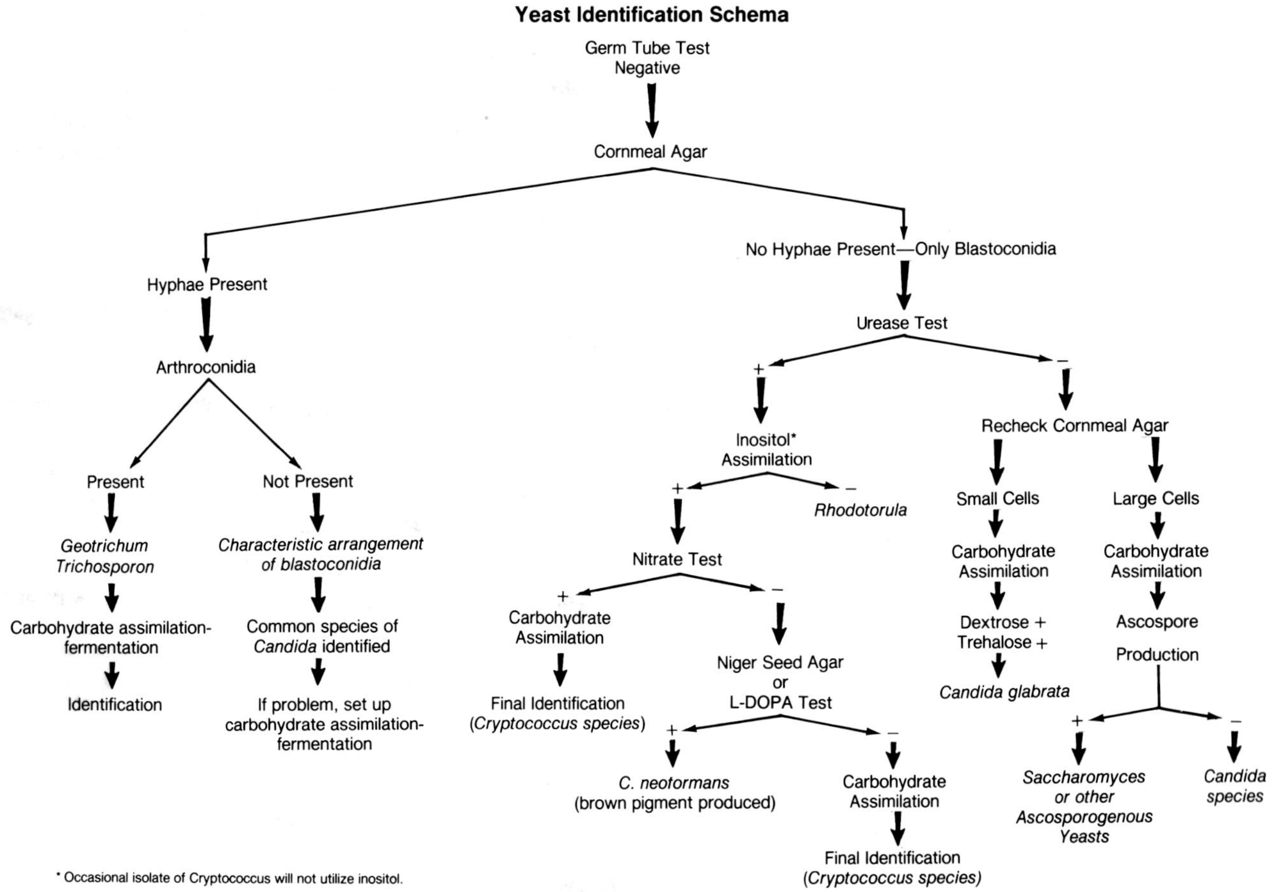


**Figure 7.2.** Schema for the practical laboratory identification of yeasts.

inadvertent lowering of the objective lens into the agar.

Observe the growth patterns and interpret results using the criteria depicted in Plate 7.1.

The formation of hyphae or pseudohyphae and the production of chlamydospores and blastoconidia in various arrangements generally provide sufficient information for the presumptive identification of *Candida* sp (Plate 7.1 *C* and *D*). Following is a brief description of the patterns formed by the more commonly encountered species:

| | |
|---|---|
| *C. albicans:* | (a) Thick-walled chlamydospores borne singly or in clusters, usually at the tips of pseudohyphae and/or, (b) blastoconidia produced in dense clusters regularly spaced along the pseudohyphae. |
| *C. tropicalis:* | Blastoconidia sparsely produced singly or in small loose clusters irregularly along the pseudohyphae. |
| *C. parapsilosis:* | Formation of spider colonies away from the agar streaks giving a sagebrush appearance. |
| *C. pseudotropicalis:* | Elongated blastoconidia arranged in parallel clusters simulating logs in a stream. |

Yeasts producing hyphae and pseudohyphae on cornmeal agar must be carefully studied for the presence of arthroconidia. If seen, *Trichosporon* sp and *Geotrichum* sp should be suspected. *Geotrichum* sp may be suspected if only arthroconidia are seen and right angle branching or germination is observed, simulating hockey sticks. *Trichosporon* sp characteristically produce both arthroconidia and blastoconidia; however, the latter are often difficult to detect. Most strains of *Trichosporon* sp are urease positive.

Some members of the genera *Candida*, *Cryptococcus*, and *Rhodotorula* do not form hyphae or pseudohyphae on cornmeal Tween-80 agar under the conditions of incubation previously described; rather, they exhibit only the production of blastoconidia. *C. glabrata* (formerly *Torulopsis glabrata*) may be suspected if uniformly sized, 2–3 micron yeasts cells with blastoconidia in compact arrangements are observed. Species of cryptococci may be suspected if yeast cells varying in size from 2–10 $\mu$M are seen. These cells are usually widely spaced due to the presence of thick polysaccharide capsules. *Rhodotorula* sp produce uniformly sized blastoconidia that give no distinctive patterns on cornmeal agar. *Rhodotorula* sp can also be suspected from the production of a red pigment or by demonstrating urease production and the lack of utilization of inosotol.

*Saccharomyces* sp commonly appear as compactly arranged cells that are relatively large and rudimentary hyphal forms may occasionally be observed.

All of these characteristics present only preliminary evidence for identification. Biochemical tests are necessary for the definitive identification of these yeasts, particularly members of the genera *Cryptococcus*, *Rhodotorula*, *Saccharomyces*, and *Candida* (see Tables 7.3 and 7.4).

## UREASE PRODUCTION

Urease production is a characteristic that is useful for the identification of cryptococci; however, *C. krusei* and members of the genera *Rhodotorula* and *Trichosporon* may also exhibit this characteristic. Two methods for the detection of urease have been previously discussed and will not be described further. An additional test, the rapid selective urease test (92), has been found useful for the definitive identification of *C. neoformans*.

**Procedure for the Rapid Selective Urease Test**

A cotton-tipped applicator impregnated with dehydrated Christensen's urease agar base is swept across the surface of 2 or 3 small colonies so that the tip is covered with the test organism.

The inoculated applicator is placed into a test tube containing 3 drops of 1% benzalkonium chloride (pH 4.86 ± 0.01) and the cotton-tipped portion is swirled firmly against the bottom of the tube to inbed the organism into the cotton fibers.

The tube is plugged with cotton and is incubated at 45° C and examined after 10, 15, 20, and 30 minutes for the presence of a color change from yellow to purple. A red color indicates urease production by *C. neoformans* (Fig. 5 of Color Plate 6).

Evaluation of this method in a recent study revealed that 99.6% of 286 isolates of *C. neoformans* yielded a positive test within 15 minutes (92). This test has been evaluated only using colonies cultured on Sabouraud's dextrose agar.

NITRATE REDUCTION

*C. neoformans* does not reduce inorganic nitrate. A rapid method described by Hopkins and Land (36) is recommended and performed as follows:

**Procedure for Rapid Nitrate Reduction Test**

The test uses a cotton-tipped applicator that has been impregnated with potassium nitrate.

The applicator is swept across 2 or 3 colonies of the yeast to be identified and is placed into a test tube where the tip is firmly pressed against the bottom to embed the cells within the cotton fibers.

The tube and applicator are incubated for 10 minutes at 45° C and after removal 2 drops of each Dimethyl-α-naphthylamine and sulfanilic acid reagents are added to the tube so that the applicator can absorb the reagents. The development of a red color indicates a positive test (Fig. 7 of Color Plate 6).

A known positive and negative control organism should be run in parallel with each test. Determination of nitrate reduction is helpful in differentiating between the species of cryptococci (see Table 7.4); however, the test is not helpful in the identification of the rhodotorulae.

L-DOPA FERRIC CITRATE TEST

*C. neoformans* produces phenoloxidase, an enzyme that can be rapidly detected by the L-DOPA ferric citrate test.

**Procedure for the L-DOPA Ferric Citrate Test**

An L-DOPA ferric citrate solution (39) is prepared (see Appendix II) and sterile blank paper disks are saturated and dried in a sterile petri dish at 37° C (disks can be stored at −20° C for up to 6 months prior to usage).

At the time a test is to be performed, an appropriate number of disks are removed from the freezer, placed into a Petri dish and moistened with phosphate buffer.

A liberal amount of the test organism is smeared onto the surface of each moistened disk.

The inoculated disks are incubated at 37° C in a sterile Petri dish and examined at 30-minute intervals for the production of a black pigment.

A positive reaction should occur within 3 minutes; the absence of pigment production is interpreted as a negative test. Both positive (*C. neoformans*) and negative (*C. albidus*) controls should be included with each test. If an unknown yeast is nitrate reductase negative and L-DOPA ferric citrate positive, one can issue a preliminary report of *C. neoformans*. Confirmation of a positive or negative result can be made by inoculating a plate of niger seed agar. Phenoloxidase is readily detected by both tests; however, the reaction on niger seed agar is considered to be the more definitive.

**Procedure for the Niger Seed Agar Test**

Plates of niger seed agar are heavily inoculated and incubated at 35° to 37° C for at least one week.

The development of a maroon red to brown to black pigment first in the area of heavy inoculum and later on the remainder of the inoculated surface indicates a positive test (Fig. 6 of Color Plate 6).

Although a positive niger seed agar test is considered to provide a definitive identification of *C. neoformans* (67), other confirmatory tests including carbohydrate utilizations are recommended to reduce the potential for error.

### CARBOHYDRATE UTILIZATION STUDIES

Carbohydrate utilization tests are perhaps the most widely used of all methods for the identification of yeasts, including the commercially available systems. All methods use a basal medium that will support growth of yeasts only if an appropriate carbohydrate source is added. Individual carbohydrate substrates may be added directly to the medium or incorporated into paper disks that are placed in contact with the medium. The presence of growth either in the medium or adjacent to the disks indicates that the carbohydrate included has been utilized by the organism being tested. In this manner, individual carbohydrate utilization patterns may be determined for yeasts and yeast-like organisms.

Several methods have been developed to determine carbohydrate utilization patterns and include: (a) a slant method developed by Adams and Cooper (2), (b) a carbohydrate-nutrient method proposed by Huppert et al. (37) and a pour plate-disk method described by Land et al. (51).

The authors of this text have found the auxonographic method described by Roberts (71) to be useful and reliable for determining carbohydrate utilization patterns.

**Procedure for the Auxanographic Method for Carbohydrate Utilization**

A suspension of the yeast to be tested is prepared, equivalent in density to a McFarland 4 standard.

With a sterile transfer pipette, the surface of a yeast nitrogen base agar plate containing bromcreosol purple is flooded with a suspension of the yeast cells. The excess inoculum is removed with a transfer pipette and the lid is left ajar for 10–15 minutes to allow the surface to dry.

Filter paper disks, 6 mm or 13 mm (Difco Laboratories, Detroit, MI, or Baltimore Biological Laboratories, Cockeysville, MD) impregnated with the appropriate carbohydrates are spaced on the agar surface, approximately 30 mm apart.

The disks are gently pressed to the surface of the agar with a pair of flamed forceps. Cultures are incubated for 24–48 hours at 30° C and are read for carbohydrate utilization.

**Table 7.3.** Characteristics of *Candida* Species and *Saccharomyces* Species Most Commonly Recovered from Clinical Specimens*

| Organism | Assimilations | | | | | | | | | | | | Fermentations | | | | | | Other Reactions | | | | |
|---|---|---|---|---|---|---|---|---|---|---|---|---|---|---|---|---|---|---|---|---|---|---|---|
| | Dextrose | Maltose | Sucrose | Lactose | Galactose | Melibiose | Cellobiose | Inositol | Xylose | Raffinose | Trehalose | Dulcitol | Dextrose | Maltose | Sucrose | Lactose | Galactose | Trehalose | Urease | $KNO_3$ | Pseudohyphae | Growth at 37° C | Germ tubes |
| *C. albicans* | + | + | + | − | + | − | − | − | + | − | + | − | F | F | − | − | F | F | − | − | + | + | + |
| *C. stellatoidea*** | + | + | − | − | + | − | − | − | + | − | +* | − | F | F | − | − | − | − | − | − | + | + | + |
| *C. glabrata* | + | − | − | − | − | − | − | − | − | − | + | − | F | − | − | − | − | F | − | − | + | + | − |
| *C. parapsilosis* | + | + | + | − | + | − | − | − | + | − | + | − | F | − | − | − | F | F | − | − | + | + | − |
| *C. tropicalis* | + | + | +* | − | + | − | +* | − | + | − | + | − | F | F | F* | − | F | F | − | − | + | + | − |
| *C. pseudotropicalis* | + | − | + | + | + | − | + | − | +* | + | − | − | F | − | F | F | F | F | − | − | + | + | − |
| *C. krusei* | + | − | − | − | − | − | − | − | − | − | − | − | F | − | − | − | − | − | +* | − | + | + | − |
| *C. lipolytica* | + | −* | − | − | − | − | − | − | − | − | − | − | F | − | − | − | − | − | − | − | −*** | − | − |
| *C. guilliermondii* | + | + | + | − | + | + | + | − | + | + | + | + | F | − | F | − | F | F | − | − | + | + | − |
| *C. rugosa* | + | − | − | − | − | + | − | − | + | − | − | − | − | − | − | − | − | − | − | − | + | − | − |
| *C. famata* | + | + | + | +* | + | +* | + | − | + | + | + | +* | F | − | F | − | − | F | − | − | − | + | − |
| *C. inconspicua* | + | − | − | − | − | − | − | − | − | − | − | − | − | − | − | − | − | − | − | − | − | + | − |
| *Saccharomyces cerevisiae* | + | + | + | − | + | − | − | − | − | + | +* | − | F | F | F | − | F | F* | − | − | + | + | + |

* Abbreviations: * = strain variation; + = assimilation; F = acid and gas; − = negative; ** = reported as *C. albicans* when *C. stellatoidea* is identified; *** = *C. lipolytica* distinguished from *C. krusei* by presence of true hyphae and pseudohyphae.

**Table 7.4.** Characteristics of *Cryptococcus* Species Most Commonly Recovered from Clinical Specimens*

| Organism | Assimilations | | | | | | | | | | | | Fermentations | | | | | | Other Reactions | | | | | |
|---|---|---|---|---|---|---|---|---|---|---|---|---|---|---|---|---|---|---|---|---|---|---|---|---|
| | Dextrose | Maltose | Sucrose | Lactose | Galactose | Melibiose | Cellobiose | Inositol | Xylose | Raffinose | Trehalose | Dulcitol | Dextrose | Maltose | Sucrose | Lactose | Galactose | Trehalose | Urease | $KNO_3$ | Pseudohyphae | Growth at 37° C | Pigment on niger seed | Capsule in India ink |
| *C. neoformans* | + | + | + | − | +* | − | + | + | + | +* | + | + | − | − | − | − | − | − | + | − | R | + | + | + |
| *C. uniguttulatus* | + | + | + | − | −* | − | −* | + | + | +* | +* | − | − | − | − | − | − | − | + | − | − | − | − | + |
| *C. albidus* var. *albidus* | + | + | + | +* | −* | −* | + | + | + | + | + | +* | − | − | − | − | − | − | + | + | +* | + | − | + |
| *C. laurentii* | + | + | + | + | + | +* | + | + | + | +* | + | + | − | − | − | − | − | − | + | − | + | + | − | + |
| *C. luteolus* | + | + | + | − | + | + | + | + | + | −* | + | +* | − | − | − | − | − | − | + | − | − | − | − | − |
| *C. albidus* var. *diffluens* | + | + | + | − | −* | +* | + | + | + | + | + | +* | − | − | − | − | − | − | + | + | −* | + | − | + |
| *C. terreus* | + | +* | −* | −* | +* | − | + | + | + | − | + | +* | − | − | − | − | − | − | + | + | −* | + | − | + |
| *C. gastricus* | + | + | −* | − | + | − | + | + | + | − | + | − | − | − | − | − | − | − | + | − | −* | − | − | + |

* Abbreviations: + = positive; R = occasional to rare hyphae; * = strain variation in assimilation; − = negative.

**Table 7.5.** Characteristics of *Trichosporon* Species and *Geotrichum* Species Most Commonly Recovered from Clinical Specimens*

| Organism | Assimilations | | | | | | | | | | | | Fermentations | | | | | | Other Reactions | | | |
|---|---|---|---|---|---|---|---|---|---|---|---|---|---|---|---|---|---|---|---|---|---|---|
| | Dextrose | Maltose | Sucrose | Lactose | Galactose | Melibiose | Cellobiose | Inositol | Xylose | Raffinose | Trehalose | Dulcitol | Dextrose | Maltose | Sucrose | Lactose | Galactose | Trehalose | Urease | $KNO_3$ | Pseudohyphae | Growth at 37° C |
| *Trichosporon beigelii* | + | +* | +* | + | + | +* | + | +* | + | +* | +* | +* | − | − | − | − | − | − | +* | − | + | + |
| *T. pullulans* | + | + | + | + | + | +* | + | +* | + | +* | + | − | − | − | − | − | − | − | + | + | + | + |
| *T. capitatum* | + | − | − | − | + | − | − | − | − | − | − | − | − | − | − | − | − | − | − | − | + | + |
| *T. inkin* | + | + | + | + | + | − | + | + | + | − | + | − | − | − | − | − | − | − | − | − | + | + |
| *Geotrichum candidum* | + | − | − | − | + | − | − | − | + | − | − | − | − | − | − | − | − | − | − | − | − | − |
| *G. penicillatum* | + | − | − | − | + | − | − | − | + | − | − | − | − | − | − | − | − | − | − | − | + | + |

* Abbreviations: * = strain variation; + = assimilation; − = negative.

A yellow color change around the carbohydrate containing disk or the presence of growth is evidence of carbohydrate utilization (Fig. 7 of Color Plate 6).

The carbohydrates most commonly used for testing hyphal- and pseudohyphal-producing yeasts include: dextrose, maltose, sucrose, lactose, and raffinose. Those used for testing nonhyphal-producing yeasts include: dextrose, maltose, sucrose, lactose, galactose, trehalose, inositol, melibiose, and raffinose (two plates must be used to accommodate this large number of disks). Carbohydrate utilization profiles for different species of yeasts may be compared with those shown in Tables 7.3 and 7.4.

### ASCOSPORE FORMATION

If *Saccharomyces* sp is suspected, a single colony may be streaked onto the surface of an ascospore agar plate. Plates are incubated at 25° C for at least 10 days. A heat-fixed smear of the organism is prepared and stained with the modified Kinyoun's acid-fast staining method. Ascospores are acid-fast and stain red when stained by the Kinyoun method (Fig. 8 of Color Plate 6). *Saccharomyces* sp or other ascosporogenous yeasts must be considered when observed.

### Summary of Practical Approach to Yeast Identification

The following step-by-step approach is useful for the identification of yeasts as presented in Table 7.1:

1. Perform a germ tube test:
   A. If positive: Report *C. albicans*
   B. If negative, go to step 2:
2. Perform a rapid urease test:
   A. If positive:
      (1) Rule out *Rhodotorula* sp by observing for the red pigmentation of the colony. *Sporobolomyces* sp may also be suspected but can be recognized by its small satellite colonies adjacent to the inoculum streak.
      (2) Set up the niger seed agar plate or an L-DOPA ferric citrate test or C/N Screen to identify *C. neoformans.* Nitrate reduction and carbohydrate utilization studies should also be performed to confirm the identification of *C. neoformans.*
   B. If negative:
      (1) Issue report of "yeast, not *C. albicans* or *C. neoformans.*" Depending upon the clinical circumstances and the origin of the culture, further species identification may be necessary. If so:
         (a) Perform carbohydrate utilization studies, using one of the commercial kit systems.
         (b) Optionally, set up a cornmeal agar plate and observe areas of growth for characteristic microscopic morphological features.
         (c) If *Saccharomyces* sp is suspected, set up a plate of ascospore agar, perform an acid-fast stain on any growth and observe for ascospores.

Each laboratory director must decide whether conventional methods or commercially available systems are appropriate for use in the laboratory. If conventional methods are used, the practical approach described in the 2nd edition of this text should be consulted for more complete instructions.

# PLATE 7.1

## MORPHOLOGY OF COMMON YEASTS AND YEAST-LIKE ORGANISMS ON CORNMEAL TWEEN-80 AGAR

*A.* Appearance of streak line on cornmeal agar. Note production of pseudohyhpae. Low power.
*B.* Growth of *C. albicans* illustrating spider-like appearance and compact clustering of blastoconidia at regular intervals along the pseudohyphae. Low power.
*C.* Detailed view of the clustering of blastoconidia of *C. albicans* placed at regular intervals along the course of the pseudohyphae. High power.
*D.* The production of pseudohyphae and terminal chlamydospores as shown here are the key identifying features of *C. albicans.* High power.
*E.* Microscopic appearance of *C. tropicalis* on cornmeal revealing relatively sparse production of blastoconidia irregularly along the pseudohyphae. High power.
*F.* Appearance of *C. parapsilosis* on cornmeal agar illustrating the characteristic spider-like satellite colonies at points distant from the streak line resulting in a sagebrush appearance. High power.
*G.* Microscopic appearance of giant hyphae of *C. parapsilosis.* High power.
*H.* Photomicrograph of *C. pseudotropicalis* on cornmeal agar illustrating the "log in stream" arrangement of the blastoconidia that is characteristic of that species. High power.

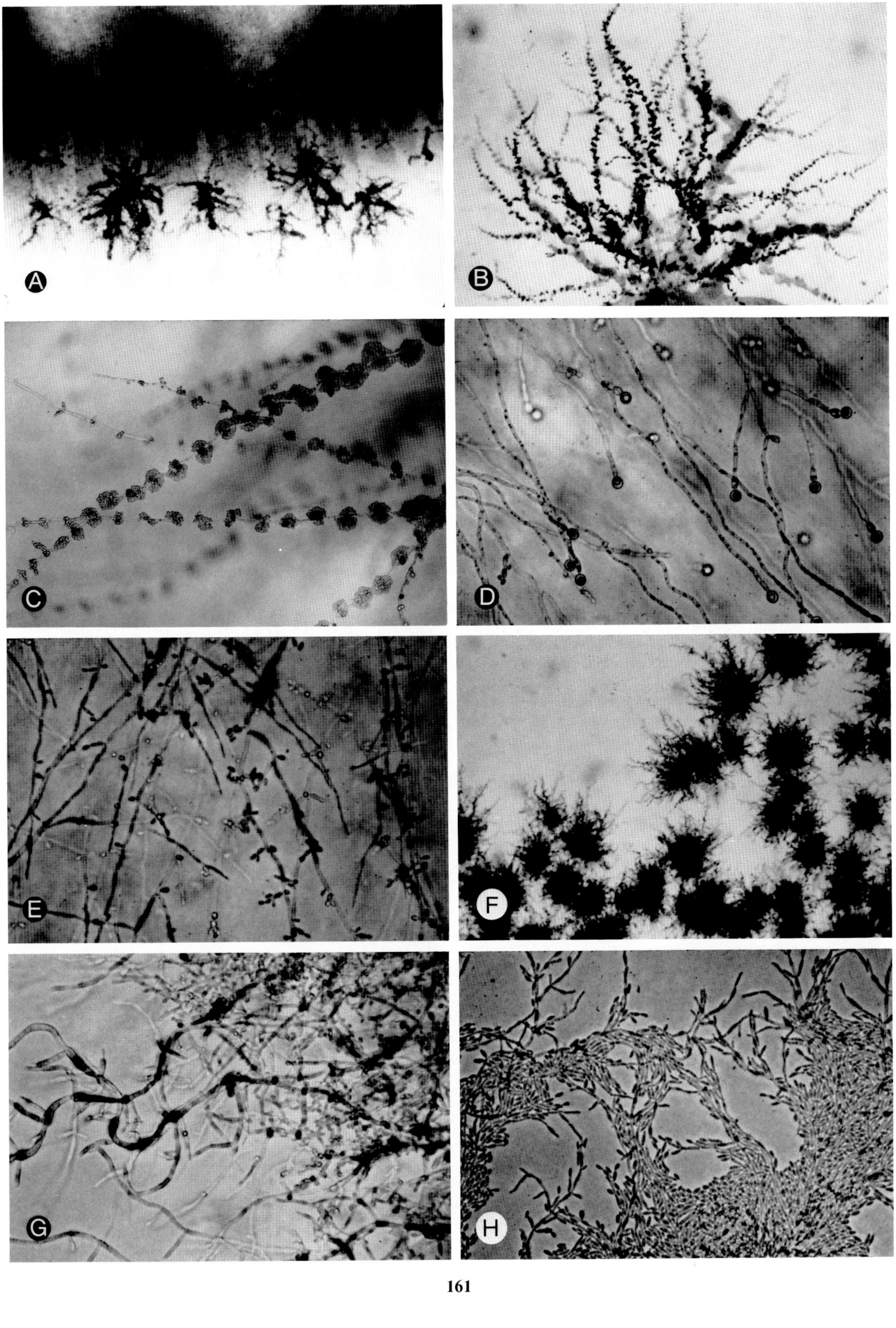
A
B
C
D
E
F
G
H

## PLATE 7.1 (*Continued*)

*I.* Appearance of *C. guilliermondii* revealing delicate, sometimes curved pseudohyphae bearing chains of ovoid blastoconidia. High power.
*J.* Branched pseudohyphae producing elongated blastoconidia and cross-stick appearance at points of septation characteristic of *C. krusei.*
*K.* Microscopic appearance of *C.* (*Torulopsis*) *glabrata* on cornmeal revealing small, spherical, highly compacted cells with no pseudohyphae present. High power.
*L.* Large, primarily spherical yeast cells characteristic of *Saccharomyces* sp. High power.
*M.* Photomicrograph of *Geotrichum* sp illustrating well developed hyphae fragmenting into arthroconidia. Note that some arthroconidia demonstrate characteristic germination from one corner. High power.
*N.* Microscopic appearance of *Trichosporon* sp illustrating hyphae with distinct arthroconidia formation. The presence of blastoconidia is the differentiating feature of *Trichosporon* species from *Geotrichum* sp when seen in cultures. High power.
*O.* *C. neoformans* on cornmeal agar showing absence of pseudohyphae. Note the separation of the irregularly sized blastoconidia by the production of capsular material. (Compare this microscopic appearance with that of *C. glabrata* in Frame *K*). Low power.
*P.* High power view of *C. neoformans* on cornmeal agar. Note spherical cells that vary in size and are widely separated by the presence of thick capsular material.

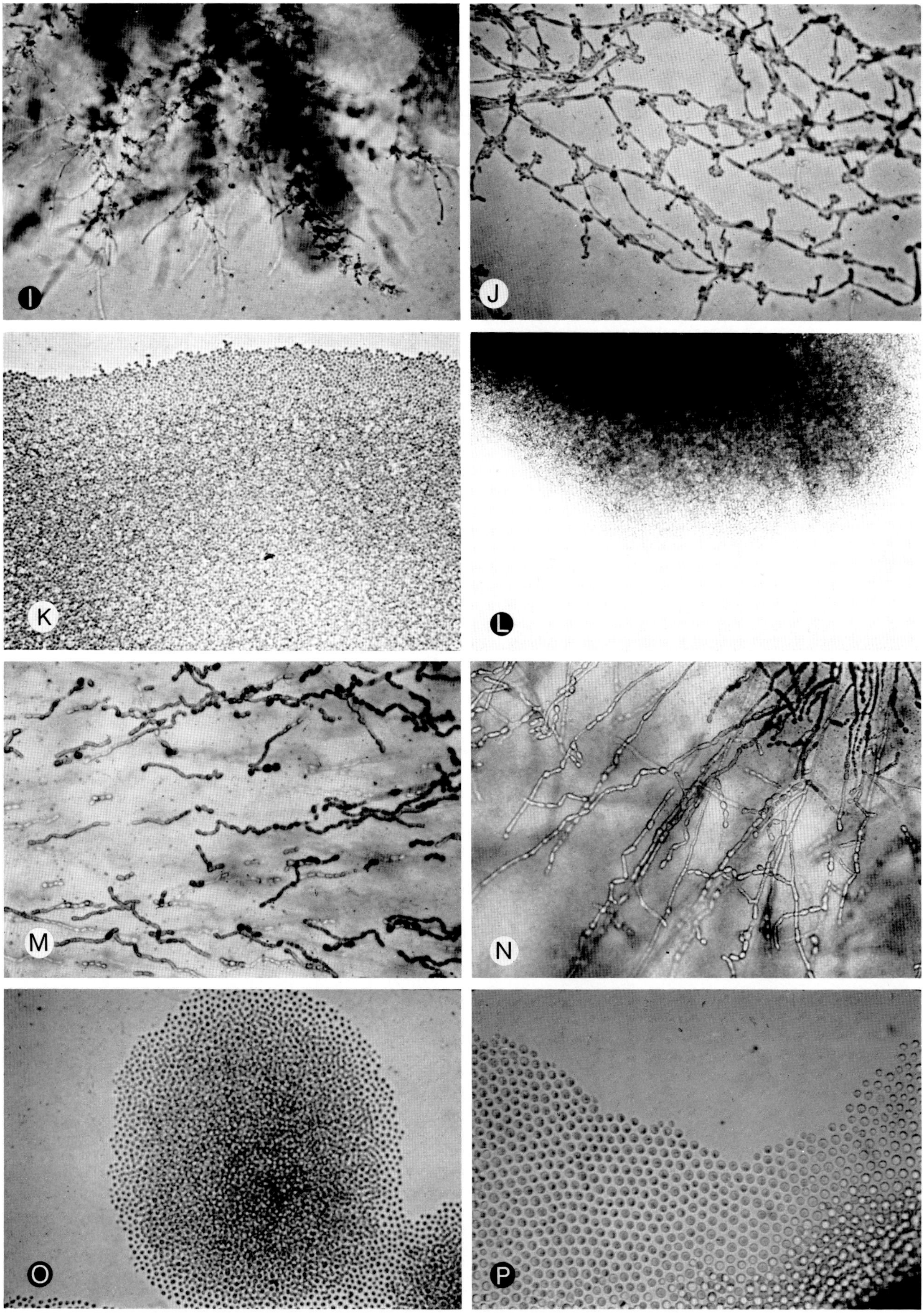
I
J
K
L
M
N
O
P

# SUGGESTED READINGS


Alexopoulos CJ and Mims CW. (1979) Introductory Mycology, 3rd ed., John Wiley & Sons, Inc., New York, London.

Barnett HL and Hunter BB. (1972) Illustrated Genera of Imperfect Fungi, 3rd ed., Burgess Publishing Co., Minneapolis.

Barron GL. (1968) The Genera of Hyphomycetes from Soil, Krieger Publishing Co, Huntington, New York.

Beneke ES and Rogers AL. (1980) Medical Mycology Manual, 4th ed., Burgess Publishing Co., Minneapolis.

Campbell MC and Steward JL. (1980) Medical Mycology Handbook. John Wiley & Sons, New York.

Carmichael JW, Kindrick WB, Conners LL, and Sigler L. (1980) Genera of Hyphomycetes, University Alberta Press, Edmonton.

Chandler FW, Kaplan W, and Ajello L. (1980) Color Atlas and Text of the Histopathology of Mycotic Diseases. Year Book Medical Publishers, Chicago.

DiSalvo AF. (1983) Occupational Mycoses, Lea & Febiger, Philadelphia.

Dolan CT, et al. (1976) Atlas of Clinical Mycology. American Society of Clinical Pathologists, Chicago.

Ellis MB. (1971) Dematiaceous Hypomycetes. Commonwealth Mycological Institute, Kew.

Emmons CW, Binford CW, Utz JP, and Kwon-Chung KJ. (1977) Medical Mycology, 3rd ed., Lea & Febiger, Philadelphia.

Haley LD and Calloway CS. (1978) Laboratory Methods in Medical Mycology. Centers for Disease Control HEW (CDC) 78-8361, Atlanta.

LaRone DH. (1976) Medically Important Fungi: A Guide to Identification. Harper & Row, Hagerstown.

Lennette EH, Balows A, Hausler WJ Jr and Truant JP. (1980) Manual of Clinical Microbiology, 3rd Ed., Chaps 52–63, American Society for Microbiology, Washington, D.C.

McGinnis MR. Laboratory Handbook of Medical Mycology. (1980) Academic Press, New York.

McGinnis MR, D'Amato RF and Land GA. (1982) Practical Handbook of Medically Important Fungi and Aerobic Actinomycetes. Praeger Scientific, New York.

Odds FC. (1979) Candida and Candidosis. University Park Press, Baltimore.

Raper KP and Fennell DI. (1965) The Genus Aspergillus. Williams & Wilkins Co., Baltimore.

Rebell G and Taplin D. (1970) Dermatophytes: Their Recognition and Identification. University of Miami Press, Coral Gables.

Rippon JW. (1982) Medical Mycology: The Pathogenic Fungi and the Pathogenic Actinomycetes, 2nd ed., W. B. Saunders, Philadelphia.

Rogers AL. (1978) Identification of Saprobic Fungi Commonly Encountered in the Clinical Environment. American Society for Microbiology, Washington, D.C.

Wilson JW and Plunkett OA. (1965) The Fungous Diseases of Man. University of California Press, Berkeley.


# REFERENCES


1. Abramowsky CR, Quinn D, Bradford WD and Conant NF: Systemic infection by *Fusarium* in a burned child. *Pediatrics* 84:561–564, 1974.
2. Adams ED Jr and Cooper BH: Evaluation of a modified Wickerham medium for identifying medically important yeasts. *Am J Med Technol* 40:377–388, 1974.
3. Agger WA and Maki DG: Mucormycosis: A complication of critical care. *Arch Intern Med* 138:925–927, 1978.
4. Aisner J, Murillo J, Schimpff SC, et al: Invasive aspergillosis in acute leukemia: Correlation with nose cultures and antibiotic use. *Ann Intern Med* 90:4–9, 1979.
5. Ayers LW, Koneman EW and Merrick TA: Surgical pathology of infectious diseases. In Silverberg, SG (ed.), Principles and Practice of Surgical Pathology, New York, John Wiley & Sons, 1983.
6. Azar P, Aquavella JV and Smith RS: Keratomycosis due to an Alternaria species. *Am J Ophthalmol* 79:881–887, 1975.
7. Beneke ES and Rogers AL: Medical Mycology Manual, 4th ed., Minneapolis, Burgess Publishing Co., 1980.
8. Beushing WJ, Kurek K and Roberts GD: Evaluation of the modified API 20C strip system for the identification of clinically important yeasts. *J Clin Microbiol* 9:565–569, 1979.

9. Bille J, Stockman L, Roberts GD, et al: Evaluation of a lysis centrifugation system for recovery of yeasts and filamentous fungi from blood. *J Clin Microbiol* 18:469–471, 1983.
10. Binford CH, Thompson K, et al: Mycotic brain abscess due to *Cladosporium trichoides*, a new species. *Am J Clin Pathol* 22:535–542, 1952.
11. Bourguignon RL, Walsh AF, Flynn JC, et al: *Fusarium* species osteomyelitis. *J Bone Joint Surg* 58A:722–723, 1976.
12. Chandler FW, Kaplan W and Ajello L: Color Atlas and Text of the Histopathology of Mycotic Diseases. Year Book Medical Publishers, Chicago, 1980.
12a. Codish SD, Sheridan ID and Monaco AP: Mycotic wound infections: A new challenge for the surgeon. *Arch Surg* 114:831–834, 1979.
13. Collins MS and Rinaldi MG: Cutaneous infection in man caused by *Fusarium moniliforme*. *Sabouraudia* 15:151–160, 1977.
14. Cooper, BH, Johnson JB and Thaxton ES: Clinical evaluation of the new Uni-Yeast-Tek system for rapid presumptive identification of medically important yeasts. *J Clin Microbiol* 7:349–355, 1978.
15. Cooper BH: Clinical laboratory evaluation of a screening medium (C/N Screen) for *Cryptococcus neoformans*. *J Clin Microbiol* 11:672–674, 1980.
16. DiSalvo AF, Flicking A, et al: *Penicillium marneffei* infection in man. Description of first natural infection. *Am J Clin Pathol* 59:259–263, 1973.
17. Dolan CT, Weed LA and Dines DE: Bronchopulmonary helminthosporiosis. *Am J Clin Pathol* 53:235–242, 1970 .
18. Drouhet EL, Martin, et al: Mycotic meningocerebrale Cephalosporin. *Presse Med* 31:1809–1814.
19. Deleted in proof.
20. Ellis MB: Dematiaceous Hyphomycetes, Kew, Surrey, England, Commonwealth Mycological Institute, 1971.
21. Emmons CW, Binford CH, Utz JP and Kwon-Chung KJ: Medical Mycology, 3rd ed., Philadelphia, Lea & Febiger, 1977, pgs. 275–277.
22. Estes SA, Merz WG and Maxwell LG: Primary cutaneous phaehyphomycosis caused by *Drechslera spicifera*. *Arch Dermatol* 113:813–815, 1977.
23. Forster RK and Rebell G: The diagnosis and management of keratomycosis. *Arch Ophthalmol* 93:975–978, 1975.
24. Farmer SG and Komorowski RA. Cutaneous microabscess formation from *Alternaria alternata*. *Am J Clin Pathol* 66:565–569, 1976.
25. Fuste FJ, Ajello L, et al: *Drechslera hawaiiensis*: Causative agent of a fatal fungal meningoencephalitis. *Sabouraudia* 11:59–63, 1973.
26. Fenech FF and Mallia CP: Pleural effusion caused by *Penicillium lilacinum*. *Br J Dis Chest* 66:284–290, 1972.
27. Garau J, Diamond RD, Lagrotteria LB, et al: *Alternaria* osteomyelitis. *Ann Intern Med* 86:747–748, 1977.
28. Green WO and Adams TE: Mycetoma in the United States. *Am J Clin Pathol* 42:75–91, 1964.
29. Grieble HG, Rippon JW, Maliwan N, et al: Scopulariopsosis and hypersensitivity pneumonitis in an addict. *Ann Intern Med* 83:326–329, 1975.
29a. Hageage GJ and Harrington BJ: Use of calcofluor white in clinical mycology. *Lab Med* 15:109–111, 1984.
30. Haldane EV, MacDonald JL, Gittens WO, et al: Prosthetic valvular endocarditis due to the fungus *Paecilomyces*. *Canad Med Assn J* 111:963–965, 1974.
31. Halde C, Padhye AA, Haley LD, et al: *Acremonium falciforme* as a cause of mycetoma in California. *Sabouraudia* 14:319–326, 1976.
32. Hall WJ: *Penicillium* endocarditis following open heart surgery and prosthetic valve insertion. *Am Heart J* 87:501–506, 1974.
33. Harris JJ and Downham TF: Unusual fungal infections associated with immunological hyporeactivity. *Int J Dermatol* 17:323–330, 1978.
34. Harris LF, Dan BM, Lefkowitz LB, et al: *Paecilomyces* cellulitis in a renal transplant patient: successful treatment with intravenous miconazole. *South Med J* 72:897–898, 1979.
35. Harris R, Smith, RE, Wood TR and Biddle M: *Helminthosporium* corneal ulcers. *Ann Ophthalmol* 20:729–733, 1978.
36. Hopkins JM and Land GA: A rapid method for determining nitrate utilization by yeasts. *J Clin Microbiol* 5:497–500, 1977.
37. Huppert M, Harper G, Sun SH and Delanerolle V: Rapid Methods for identification of yeasts. *J Clin Microbiol* 2:21–34, 1975.
38. Kamalam A and Thambiah AS: Cutaneous infection by *Syncephalastrum*. *Sabouraudia* 18:19–20, 1980.
39. Kaufman CS and Merz WG: Two rapid pigmentation tests for identification of *Cryptococcus neoformans*. *J Clin Microbiol* 15:339–341, 1982.
40. Kaufman SM: *Curvularia* endocarditis following cardiac surgery. *Am J Clin Pathol* 56:466–470, 1971.
41. Keys TF, Haldorsen AM, Rhodes KH, et al: Nosocomial outbreak of *Rhizozopus* infections associated with elastoplast wound dressings—Minnesota. *Morbidity Mortality Weekly Report* 27:33–34, 1978.
42. Kiehn TE, Edwards FF and Armstrong D: The prevalence of yeasts in clinical specimens from cancer patients. *Am J Clin Pathol* 73:518–521, 1980.
43. Kirkpatrick MB, Pollock HM, Wimberley NE, et al: An intracavitary fungus ball composed of *Syncephalastrum*. *Am Rev Respir Dis* 120:943–947, 179.
44. Koneman EW and Roberts GD: Clinical and laboratory diagnosis of mycotic disease. In Henry, J. B. (ed): Clinical Diagnosis and Management by Laboratory Methods, 17th Edition, Philadel-

phia, W. B. Saunders Co., 1984.

45. Krachmer JH, Anderson RL, Binder PS, et al: *Helminthosporium* corneal ulcers. *Am J Ophthalmol* 85:666–670, 1978.
46. Kurung, JM: The isolation of *Histoplasma capsulatum* from sputum. *Am Rev Tuberc* 66:578, 1952.
47. Kwong-Chung KJ, Schwartz IS and Rybak BJ: A pulmonary fungus ball produced by *Cladosporium cladosporioides. Am J Clin Pathol* 64:564–568, 1975.
48. Kwon-Chung KJ, Young RC, Orlando M: Pulmonary mucormycosis caused by *Cunninghamella elegans* in a patient with chronic myelogenous leukemia. *Am J Clin Pathol* 64:544–548, 1975.
49. Laham MN and Carpenter JL: *Aspergillus terreus*, a pathogen capable of causing infective endocarditis, pulmonary mycetoma and allergic bronchopulmonary aspergillosis. *Am Rev Respir Dis* 125:769–772, 1982.
50. Lampert RP, Hutto JH, et al: Pulmonary and cerebral mycetoma caused by *Curvularia pallescens. J Pediatr* 91:603–605, 1977.
51. Land GA, Vinton EC, Adcock GB and Hopkins JM: Improved auxanographic method for yeast assimilation: A comparison with other approaches. *J Clin Microbiol* 2:2106–2107, 1975.
52. Land GA, Harison BA, Hulme KL, Cooper BH and Byrd JC: Evaluation of the new API 20C strip for yeast identification against a conventional method. *J Clin Microbiol* 10:357–364, 1979.
53. Liebler GA, Magovern GJ, Sadighi P, et al: *Penicillium* granuloma of the lung presenting as a solitary pulmonary nodule. *JAMA* 237:671, 1977.
54. Lobritz RW, Roberts TH, Marraro RV, et al: Granulomatous pulmonary disease secondary to *Alternaria. JAMA* 241:596–597, 1979.
55. Loveless MO, Winn RE, Campbell M, et al: Mixed invasive infection with *Alternaria* species and *Curvularia* species. *Am J Clin Pathol* 76:491–493, 1981.
56. MacKinnon JE: Mycetomas as opportunistic wound infections. *Lab Invest* 11:1124–1131, 1962.
57. McGinnis MR: Human pathogenic species of *Exophiala, Phialophora* and *Wangiella.* Proc IV Int Conf Mycoses P.A.H.O. Sci. Pub. 356:37–59, 1978.
58. McGinnis MR and Schell WA: The genus *Fonsecaea* and its relationship to the genera *Cladosporium, Phialophora, Ramichloridium* and *Rhinocladiella.* Proc Vth Int Conf Mycoses P.A.H.O. Sci. Pub. 396:215–234, 1980.
59. Middleton FG, Jurgenson PF et al: Isolation of *Cladosporium trichoides* from nature. *Mycopathologia*, 62:125–127, 1976.
60. Meyers BR, Wormser G, Hirschman SZ, et al: Rhinocerebral mucormycosis: Premortem diagnosis and therapy. *Arch Intern Med* 139:557–560, 1979.
61. Mosier MA, Lusk B, Pettit RH, et al: Fungal endophthalmitis following intraocular lens implantation. *Am J Ophthalmol* 83:1–8, 1977.
62. Murray PR, VanScoy RE and Roberts GD: Should yeast in respiratory secretions be identified? *Mayo Clin Proc* 52:42–45, 1977.
63. Nityananda K, Sivasubramaniam P, Ajello L: Mycotic keratitis caused by *Curvularia lunata*: Case report. *Sabouraudia* 2:35–36, 1962.
64. Odds FC: Candida and Candidosis. Baltimore, University Park Press, 1979.
65. Onorato IM, Axelrod JL, et al: Fungal infections of dialysis fluid. *Ann Int Med* 91:50–52, 1978.
66. O'sullivan FX, Stuewe BR, Lynch JM, et al: Peritonitis due to *Drechslera spicifera* complicating continuous ambulatory peritoneal dialysis. *Ann Int Med* 94:213–214, 1981.
67. Paliwal DK and Randhawa HS: Evaluation of a simplified *Guizotia abyssinicia* seed medium for differentiation of *Cryptococcus neoformans. J Clin Microbiol* 7:346–348, 1978.
68. Philip A: Some infrequently isolated fungi in a clinical laboratory. *Lab Med* 14:158–162, 1983.
69. Raper KB and Fennel DI: The Genus *Aspergillus*, Baltimore, Williams & Wilkins, 1965.
70. Rippon JW: Medical Mycology—The Pathogenic Fungi and the Pathogenic Actinomycetes, Philadelphia, WB Saunders Co., 1982, pg. 266.
71. Roberts GD: Laboratory diagnosis of fungal infections. *Hum Pathol* 7:161–168, 1976.
72. Roberts GD, Karlson AG and DeYoung DR: Recovery of pathogenic fungi from clinical specimens submitted for mycobacterial culture. *J Clin Microbiol* 3:47–48, 1976.
73. Roberts GD, Horstmeier CD, Foxworth JH and Land GA: Rapid urea broth test for yeasts. *J Clin Microbiol* 7:584–588, 1978.
74. Roberts GD: "Laboratory Diagnosis of Mycotic Infections." Unpublished observation. See also: Hariri AR, Hempel HO, Kimberlin CL, et al: Effects of time lapse between sputum collection and isolation of clinically significant fungi. *J Clin Microbiol* 15:425–428, 1982.
75. Rockhill RC and Klein MD: *Paecilomyces lilacinus* as the cause of chronic maxillary sinusitis. *J Clin Microbiol* 11:737–739, 1980.
76. Rodriques MM and MacLeod D: Exogenous fungal endophthalmitis caused by *Paecilomyces. Am J Ophthalmol* 79:687–690, 1975.
77. Rohwedder JJ, Simmons JL, et al: Disseminated *Curvularia lunata* infection in a football player. *Arch Intern Med* 139:940–941, 1979.
78. Rowsey JJ, Acers TE, Smith DL, et al: *Fusarium oxysporum*

endophthalmitis. *Arch. Ophthalmol* 97:103–105, 1979.

79. Rush-Munro FM, Black H and Dingley JM: Onychomycosis caused by *Fusarium oxysporum. Aust J Dermatol* 12:18–29, 1971.
80. Seabury JH and Samules M: The pathogenic spectrum of aspergillosis. *Am J Clin Pathol* 40:21–33, 1963.
81. Seligsohn R, Rippon JW and Lerner SA: *Aspergillus terreus* osteomyelitis. *Arch Intern Med* 137:918–920, 1977.
82. Sekhon AS, Willans DJ and Harvey JH: Deep scopulariopsosis: a case report and sensitivity studies. *J Clin Pathol* 27:837–843, 1974.

82a. Smith CD, Goodman NL: Improved culture medium for isolation of *Histoplasma capsulatum* and *Blastomyces dermatitidis. Am J Clin Pathol* 63:276–280, 1975.

83. Stuart EA and Blank F: Aspergillosis of the ear. A report of twenty-nine cases. *Can Med Assoc J* 72:334–337, 1955.
84. Takayasu S, Akagi M and Shimizu Y: Cutaneous mycosis caused by *Paecilomyces lilacinus. Arch Dermatol* 114:1687–1690, 1977.
85. Thompson RB Jr and Roberts GD: A practical approach to the diagnosis of fungal infections of the respiratory tract. *Clinics in Lab Med* 2:321–342, 1982.
86. Vermeil CA, Gordeff A, et al: *Blastomyces cheloidiinne a Aureobasidium pullulans. Mycopathologia*, 43:35–39, 1971.
87. Wheeler S, McGinnis MR, Schell WA, et al: *Fusarium* infection in burned patients. *Am J Clin Pathol* 75:304–311, 1981.
88. Young CN, Swart JG, Ackermann D, et al: Nasal obstruction and bone erosion caused by *Drechslera hawaiiensis. J Laryngol Otol* 92:137, 1978.
89. Young MA, Kwon-Chung, KJ, Kubota TT, et al: Disseminated infection by *Fusarium moniliforme* during treatment for malignant lymphoma. *J Clin Microbiol* 7:589–594, 1978.
90. Zapater RC and Arrechea A: Mycotic keratitis by *Fusarium. Ophthalmologia*, 170:1–12, 1975.
91. Zapater RC, Albesi EJ and Garcia GH: Mycotic keratitis by *Drechslera spicifera. Sabouraudia*, 13:295–298, 1975.
92. Zimmer BL and Roberts GD: Rapid selective urease test for presumptive identification of *Crytococcus neoformans. J Clin Microbiol* 10:380–381, 1979.

# APPENDIX I

# APPENDIX I

# Glossary of Useful Mycological Terms

**Abscess:** Localized collection of pus in cavity formed by disintegration of tissue.

**Abstriction:** Formation of spores by the cutting off of successive portions of the sporophore through the growth of septa.

**Acrogenous:** Borne at the tip. Applied to spores that develop at the tip of a conidiophore.

**Acropetal:** Asexual spore produced by successive budding of the distal spore in a spore chain. The youngest spore in the chain is at the tip.

**Acropleurogenous:** Bearing spores at the tip and on the sides of the mycelium.

**Acuminate:** Gradually tapering to a point.

**Adenopathy:** Any disease of the glands, especially the lymph glands.

**Aerial:** Growing, forming, or existing in the air.

**Aerobic:** Requiring the presence of oxygen to grow.

**Aleuriospore:** Nondecidious spore which fractures the wall of the hypha to which it was attached when it is broken off.

**Algae:** Any one of several phyla of the plant kingdom, including those microorganisms containing chlorophyll but not leaf-like and stem-like parts.

**Amerospore:** Single-celled asexual spore, i.e., aseptate.

**Anaerobic:** Living in the absence of oxygen.

**Annellide:** The ridge or scar produced on a sporogenous cell at the site of derivation of an expanding conidiogenous cell.

**Anthropophilic:** Term applied to fungi that usually infect man only.

**Apical:** Located at the tip of a pointed extremity. In mycology, generally refers to sporulation from the tip of a specialized hyphal structure.

**Arciform:** Shaped like an arch or bow.

**Arthroconidium:** Asexual spore formed by the disarticulation of the mycelium.

**Ascocarp:** General term for a mycelial sac within which are formed asci and ascospores.

**Ascomycota:** Large division of higher fungi distinguished by septate hyphae and by sexual spores formed in asci or spore sacs.

**Ascospore:** Sexual spore characteristic of the *Ascomycota*, produced in a sac-like structure, known as an ascus, after the union of two nuclei.

**Ascus:** Sack-like structure containing (usually eight) ascospores developed during sexual reproduction.

**Aseptate:** Lacking cross walls.

**Basidiomycota:** Large division of fungi distinguished by septate hyphae, often large, fruiting bodies, and spores borne on a characteristic clublike basidium (mushrooms, puffballs, smuts, and rusts).

**Basidiospore:** Sexual spore characteristic of the *Basidiomycota*, produced after the union of two nuclei on a specialized club-like structure known as the basidium.

**Basidium:** Club-shaped specialized cell of the *Basidiomycota* on which are borne the exogenous basidiospores.

**Basipetal:** Production of an asexual spore produced by successive budding of the basal spore in the spore chain. The apical spore is the oldest.

**Blastoconidium:** Spore produced by a budding process along the mycelium or by a single spore.

**Budding:** Asexual reproductive process characteristic of unicellular fungi or spores involving the formation of lateral outgrowths from the parent cells that are pinched off to form new cells.

**Capitate:** Referring to a hemispherical colony.

**Capsule:** Hyaline, mucopolysaccharide sheath on the wall of a cell or spore.

**Carotenoid pigment:** Red or yellow pigment containing varying proportions of one or more carotenoid compounds.

**Catenulate:** Occurring in a chain-like or linear arrangement.

**Cerebriform:** Possessing brain-like folds.

**Chancre:** Punched-out ulcer; most commonly refers to the painless ulcer of primary syphilis.

**Chlamydospore:** Thick walled, resistant spore formed by the direct differentiation of the mycelium in which there is a concentration of protoplasm and nutrient material.

**Clavate:** Club-shaped.

**Cleistothecium:** Sexual structure, usually spherical, in which asci are contained and ascospores are formed.

**Coenocytic:** Term applied to a cell or an aseptate hypha containing numerous nuclei.

**Columella:** Sterile, inflated end of a sporangiophore, extending into a sporangium.

**Conidia:** Asexual fungal spores which are abstricted in various ways from the conidiophore.

**Conidiophore:** Specialized aerial hypha bearing conidia.

**Coremium:** Fruiting body consisting of a sterile stalk of parallel hyphae and a terminal head of fertile or spore-bearing branches and twisted strands, each composed of many hyphal threads, giving the colony the gross appearance of a prickly pear or cactus.

**Deciduous:** Falling off when ripe.

**Dematiaceous:** Dark, referring to the dark or black fungi.

**Denticle:** A peg-like projection to which conidia are sometimes attached.

**Dermatitis:** Inflammation of the skin.

**Dermatophyte:** Fungus living as a parasite on skin, hair, or nails of man or animals.

**Deuteromycota:** Large group of fungi in which the asexual stage of reproduction is known, but not the sexual stage.

**Dichotomous:** Branching in two directions, e.g., 45° branching.

**Dictyospore:** Multicelled asexual spore divided by septa in two or more planes.

**Didymospore:** Two-celled asexual spore.

**Dimorphic:** Having two forms. Refering to the mold/yeast forms of certain pathogenic fungi.

**Disseminated:** Disposed in separate patches. Referring to spread of fungus infection to two or more body sites.

**Echinulate:** Spiny.

**Ectothrix:** Outside the hair shaft.

**Endemic:** Confined or indigenous to a certain area or region.

**Endogenous:** Produced or originating from within.

**Endospore:** Asexual spore developed within a cell.

**Endothrix:** Within the hair shaft.

**Exogenous:** Produced or originating from without.

**Favic chandeliers:** Specialized hyphae that are curved, freely branching, and antler-like in appearance. Found in certain dermatophytes, especially *Trichophyton schoenleinii.*

**Faviform:** Convoluted or honeycomb-like appearance in some colonies.

**Favus:** Contagious tinea capitis caused by certain dermatophytes, notably *Trichophyton schoenleinii* and *T. violaceum.*

**Filament:** Long, cylindrical, thread-like, single cell.

**Fission:** Division of a cell into two cells by splitting.

**Fistula:** Deep sinus-forming ulcer, commonly communicating between a hollow visceral organ and the skin or mucous membrane.

**Floccose:** Woolly appearance on a colony surface.

**Fragmentation:** Breaking or segmenting of the hyphae

into fragments, each of which is capable of forming a new organism.

**Fungi imperfecti:** Large group of fungi in which the asexual stage of reproduction is known, but not the sexual stage.

**Fusiform:** Spindle-shaped.

**Geniculate:** Bent like a knee, a knot-like structure.

**Genus:** Division of a family which contains related species.

**Geophilic:** Term applied to fungi whose natural habitat is in the soil.

**Germ tube:** Tube-like process put out by a germinating spore that develops into the mycelium.

**Glabrous:** Smooth.

**Gummatous:** In the nature of a soft, gummy tumor.

**Heterotrophic:** Using organic compounds as energy sources.

**Host:** Any plant or animal that supports a parasite.

**Hyaline:** Colorless, transparent.

**Hyphae:** Filaments that make up the thallus or body of a fungus.

**Imperfect state:** Asexual stage; phase of life cycle in which there is no sexual reproduction.

**Intercalary:** Said of spores produced between two cells, e.g., chlamydospore in center of a hypha.

**Internodes:** Those areas on the stolon between the points where rhizoids are subtended.

**Intertriginous:** Affected with or of the nature of intertrigo; dermatitis occurring between two folds of the skin.

**Kerion:** Pustular disease of the scalp in which infection produces a boggy lesion.

**Lateral:** Said of spores produced on the side of the hyphae.

**Macroconidium:** Larger of two types of conidia in fungi.

**Merosporangium:** Cylindrical outgrowth from swollen end of a sporangiophore in which a chain-like series of sporangiospores is generally produced.

**Metulae:** Secondary branches of conidiophores of *Penicillium* that support the phialides.

**Microconidium:** Smaller of two types of conidia in fungi.

**Mucilaginous:** Slimy and adhesive.

**Multiseptate:** Having multiple dividing walls or partitions.

**Muriform:** Term applied to multicelled conidia divided by both vertical and horizontal septa.

**Mycelium:** Mat of intertwined and branching hyphae.

**Mycetoma:** Madura foot or fungus foot; a disease marked by swelling of the foot, in which nodules develop, followed by collection of pus and sinus formation.

**Myxomycetes:** Class of peculiar organisms, the slime molds.

**Nodal:** A term most commonly used to indicate the derivation of sporangiophores adjacent to the rhizoids in *Rhizopus* sp.

**Nodular body:** One or more closely intertwined hyphae forming a rounded, ball-like structure.

**Nonseptate:** Lacking septa (coenocytic).

**Onychia:** Ulceration of matrix of a nail.

**Onychomycosis:** Mycotic infection of the nails produced by any of a number of fungi.

**Oospore:** Sexual spore produced through the fusion of two unlike gametangia; found in the Chytridiomycota.

**Otomycosis:** Disease of the external ear caused by various fungi.

**Parasite:** Organism that lives upon and gets its food from another living organism.

**Paronychia:** Inflammation or ulceration of tissue around the nail.

**Pathogen:** Any disease-producing microorganism.

**Pectinate bodies:** Vegetative mycelial branches with unilateral digitate projections resembling teeth of a comb.

**Pedicel:** Any slender stalk, especially one that supports a fruiting or spore-bearing organ.

**Pellicle:** Thin film or scum formed on the surface of a liquid medium by the growth of a fungus.

**Penicillus:** Little brush.

**Perfect stage:** Sexual stage; stage of life cycle in which spores are formed after nuclear fusion.

**Perithecium:** Special round, oval, or beaked structure, having a small opening, within which asci are formed.

**Phaeohyphomycosis:** A subcutaneous abscess caused by one of the filamentous dark fungi.

**Phialide:** Specialized portion of the conidiophores, usually flask-shaped, from which arise the conidia.

**Pleomorphism:** Degenerative change in a fungus that converts the colony into one that is completely sterile. The characteristic and diagnostic spores required for identification are lost and cannot be recovered; usually pleomorphism is irreversible.

**Pleurogenous:** Borne on sides of conidiophore or hypha.

**Pseudohyphae:** Not true hyphae; usually refers to elongated blastoconidia formed by budding yeasts.

**Pseudomycelium:** Loosely united catenulate groups of cells formed by apical budding which, when elongated, resemble mycelial hyphae.

**Pulvonate:** Refers to a colony that is distinctly convex.

**Pycnidium:** Asexual, globose, or flask-like fruiting body containing conidia.

**Pyriform:** Pear-shaped.

**Racquet hypha:** Enlarged, club-shaped hypha with the smaller end attached to the large end of an adjacent club-shaped hypha.

**Rhizoid:** Radiating hyphae, branched and root-like, extending into the substrate.

**Ringworm:** Term used to designate superficial fungus infections; derived from the ancient belief that these infections were caused by worm-like organisms. Also derived from the circinate or circular form of the lesions.

**Rugose:** Wrinkled or folded.

**Saprophyte:** Any plant organism that obtains nourishment from dead organic matter.

**Schizomycetes:** Order of organisms lacking chlorophyll; the bacteria.

**Sclerotium:** Hard, compact mass of mycelium; the resting stage of certain fungi.

**Scutulum:** Crust-like lesion seen in favus.

**Septate:** Divided by cross walls.

**Septum:** Cross wall in a hyphal filament.

**Serpiginous:** Creeping from part to part.

**Sessile:** Attached directly by the base, without a stalk.

**Simple:** Unbranched.

**Species:** Category of classification lower than a genus or subgenus and above a subspecies or variety; its members possess certain characteristics in common.

**Spherule:** A thick-walled, closed, usually spherical hyphal structure enclosing asexual endospores. The tissue form of *Coccidioides immitis* and *Rhinosporidium seeberi.*

**Spiral hyphae:** Coiled or corkscrew-like turns in hyphae.

**Sporangiophore:** Specialized mycelial branch bearing a sporangium.

**Sporangiospore:** A spore borne within a sporangium.

**Sporangium:** Closed structure within which asexual spores are produced by cleavage.

**Spore:** Small reproductive unit or body, functioning like a seed.

**Sporodochium:** Cushion-shaped aggregate of conidiophores.

**Sterigmata:** Short or elongate specialized projections from sporophores on which spores are developed.

**Stolon:** Runner. A horizontal hypha which sprouts where it touches the substrate and forms rhizoids in the substrate.

**Stroma:** Cushion-like mat of fungal cells.

**Subcutaneous:** Situated or occurring underneath the skin.

**Symbiosis:** Living together of unlike organisms, to the benefit of each.

**Sympodial:** Growth of a conidiophore after the main axis has produced a terminal conidium. The growth appears twisted (bent-knee-like) and conidia are produced below and to one side of the previous conidium.

**Synnemata:** Stiffly cemented cluster of elongated conidiophores.

**Terminal:** Location of a spore at the tip of a hyphal segment. Most commonly used to describe the position of chlamydospores on the pseudohyphae of *Candida albicans.*

**Thallospore:** Spore derived from a vegetative cell of the thallus. Term antiquated.

**Thallus:** Term used for the fungus plant. Term antiquated.

**Tinea (ringworm):** Prefix used to designate various types of superficial fungus infections. Examples: Tinea capitis, ringworm of the scalp; tinea pedis, ringworm of the foot.

**Tuberculate:** Covered with knob-like excrescences.

**Umbilicate:** Having a depressed center.

**Umbonate:** Having a knob or elevation at or near the center of a colony.

**Vegetative:** Referring to hyphae that extend into the substrate and are the food-absorbing portion of a fungus.

**Verrucose:** Covered with wart-like projections.

**Versicolor:** Having various colors.

**Vesicle:** Blister or bladder-like structure.

**Virulence:** Degree of pathogenicity.

**Wood's lamp:** Apparatus producing filtered ultraviolet rays.

**Zoophilic:** Term applied to fungi that infect lower animals as well as humans.

**Zoospore:** Motile asexual spore.

**Zygospore:** Thick-walled sexual spore produced through fusion of two similar gametangia; found in the Zygomycota.

# APPENDIX II

# APPENDIX II

# Media, Stains and Reagents

## CULTURE MEDIA

### ASCOSPORE AGAR

| | |
|---|---|
| Potassium acetate | 10.0 g |
| Yeast extract | 2.5 g |
| Dextrose | 1.0 g |
| Agar | 30.0 g |
| Distilled water | 1000.0 ml |

Boil to put into solution. Sterilize 15 psi/15 minutes. Pour into sterile, plastic Petri dishes (15 ml/plate).

***Purpose***

For production of ascospores in ascosporogenous yeasts, e.g., *Saccharomyces.*

### BENNETT'S AGAR

| | |
|---|---|
| Yeast extract | 1 g |
| Beef extract | 1 g |
| N-Z amine A* | 2 g |
| Dextrose | 10 g |
| Agar | 15 g |
| Distilled water | 1000 ml |

Stir until dissolved, heat to boiling (stir constantly). Autoclave 15 psi/15 minutes. Final pH is 7.3. Dispense into sterile plastic Petri dishes (35 ml/plate).

***Purpose***

For cultivation of *Nocardia* and *Streptomyces.*

### BLOOD AGAR WITH CHLORAMPHENICOL AND GENTAMICIN

Add 52 g of the commercially prepared brain-heart infusion agar (Difco) to 1 liter of distilled water. Boil to dissolve and add gentamicin to make a final concentration of 5 $\mu$g/ml, and chloramphenicol to make a final concentration of 16 $\mu$g/ml.

---

* Sheffield Chemical, Division National Dairy Products Corporation, Norwich, New York.

Chloramphenicol is dissolved in 2 ml of 95% ethanol and is then added to the medium. Autoclave for 15 minutes at 121 C at 15 psi. When medium has cooled, add sheep blood to make a final concentration of 10%. Final pH is 7.4. Plates should contain 35–40 ml of medium each.

***Purpose***

For isolation of fungi excluding dermatophytes.

## BLOOD AGAR WITH CHLORAMPHENICOL, GENTAMICIN, AND CYCLOHEXIMIDE

Add 52 g of commercially prepared brain-heart infusion agar (Difco) to 1 liter of distilled water. Boil to dissolve and add the following antibiotics:

1. Gentamicin, to give a final concentration of 5 μg/ml.
2. Chloramphenicol, to give a final concentration of 16 μg/ml (dissolve in 2 ml of 95% ethanol before adding to medium).
3. Cycloheximide* (Upjohn), to give a final concentration of 500 μg/ml (dissolve in 5 ml of acetone before adding to medium).

Autoclave for 15 minutes at 121 C at 15 psi. When medium has cooled, add sheep blood to give a final concentration of 10%. Final pH is 7.4. Each plate should contain 35 to 40 ml of medium.

***Purpose***

For isolation of fungi excluding dermatophytes. Saprobes will be inhibited.

## BRAIN-HEART INFUSION AGAR (DIFCO)

Formula (ingredients per liter):

| | |
|---|---|
| Calf brain infusion | 200.0 g |
| Beef heart infusion | 250.0 g |
| Proteose-peptone | 10.0 g |
| Bacto-dextrose | 2.0 g |
| NaCl | 5.0 g |
| Disodium phosphate | 2.5 g |
| Agar | 15.0 g |

Final pH is 7.4 ± at 25 C.

* Cycloheximide is available as Actidione from the Upjohn Company, Kalamazoo, Michigan.

Add 52 g of the commercially prepared media and to 1 liter of distilled water. Boil to dissolve and autoclave at 15 psi/15 minutes. Pour into sterile, plastic Petri dishes (35 ml/plate).

***Purpose***

For isolation and subcultures of fungi and *Nocardia.*

## BRAIN-HEART INFUSION AGAR

### Biphasic Blood Culture Bottles

| | |
|---|---|
| Brain-heart infusion (Difco) | 37 g |
| Agar | 17 g |
| Distilled water | 1000 ml |

Boil the above mixture to disperse the agar. Pour into 8-oz prescription bottles, 50 ml/bottle. Place an antigen bottle rubber stopper on the bottle and autoclave for 15 min at 121 C at 15 psi. Slant the agar in the bottles by resting them on their flat side on a level bench. Allow to cool for 24 hours.

Make a solution of brain-heart infusion broth by dissolving 37 g of brain-heart infusion in 1000 ml of distilled water. Sterilize for 15 minutes at 121 C at 15 psi. Add 60 ml to each of the bottles containing agar.

Addition of the broth may be best accomplished through a sterile hypodermic needle and rubber tubing. Incubate the bottles for 24 hours at 37° C to check for sterility.

***Purpose***

For the recovery of fungi from blood.

## CASEIN AGAR

| | |
|---|---|
| Skim milk (dehydrated or instant non-fat milk) | 50 g |
| Agar | 10 g |
| Distilled water | 1000 ml |

Dissolve skim milk in 500 ml of distilled water, stir until dissolved and heat; but do not boil.

Dissolve 10 g of agar in 500 ml of distilled water and heat to boiling.

Autoclave each solution separately for 15 minutes at 15 psi.

Cool both solutions to 45° C, mix together and pour 20 ml of each into sterile plastic Petri dishes.

***Purpose***

For identification of *Nocardia* and *Streptomyces.*

## CORNMEAL AGAR

| | |
|---|---|
| Distilled water | 1500.0 ml |
| Cornmeal (yellow) | 62.5 g |

Heat in a water bath for 1 hour at 52° C. Filter through filter paper and make up the volume of the filtrate to 1500 ml with distilled water.

Add 19 g of agar. Boil to dissolve. Approximately 175 ml can be added to 8-oz bottles for storage if desired.

Autoclave 15 psi/15 minutes at 121° C.

Dispense into sterile, plastic Petri dishes from the above-mentioned storage bottles. The agar in the storage bottle can be melted in a boiling water bath until liquid.

***Purpose***

For use in identification of dermatophytes.

## CORNMEAL AGAR WITH TWEEN 80 AND TRYPAN BLUE

Use plain cornmeal agar prepared as described above.
Add 1.5 ml of trypan blue (1% solution) and 15 ml of Tween 80 **after** boiling.
Sterilize 15 psi/15 minutes.
Pour into sterile, plastic Petri dishes (15 ml/plate).

### Trypan Blue (1% Solution)

| | |
|---|---|
| Trypan blue | 0.1 g |
| Distilled water | 10.0 ml |

### Tween 80

| | |
|---|---|
| Tween 80 | 200.0 ml |
| Distilled water | 800.0 ml |

***Purpose***

For morphological identification of yeasts.

## COTTONSEED CONVERSION MEDIUM

| | |
|---|---|
| Dextrose | 20 g |
| Agar | 10 g |

| | |
|---|---|
| Pharmamedia* | 20 g |
| Distilled water | 1000 ml |

Heat to disperse the agar. Dispense into sterile tubes (25 × 100 ml) 10 ml/tube. Autoclave 15 psi/15 minutes. Slant the tubes and allow to cool.

***Purpose***

For conversion of filamentous form of *Blastomyces dermatitidis* to the characteristic yeast form.

### CZAPEK DOX AGAR (BBL)

| | |
|---|---|
| Sucrose | 30 g |
| Sodium nitrate | 2 g |
| Dipotassium phosphate | 1 g |
| Magnesium sulfate | 0.5 g |
| Potassium chloride | 0.5 g |
| Ferrous sulfate | 0.01 g |
| Agar | 15 g |
| Distilled water | 1000 ml |

Final pH is 7.3. Mix the above and heat to boiling to dissolve completely. Autoclave 15 psi/15 minutes.

A broth solution can be made using the same ingredients listed above with the omission of the agar.

Dispense the agar solution into sterile, plastic Petri dishes 35 ml/plate. Broth solutions are dispensed into sterile bottles, 100 ml/bottle.

***Purpose***

For species identification of aspergilli.

### ENRICHED CYSTINE HEART AGAR (BBL)

| | |
|---|---|
| Beef heart infusion | 500 g/liter |
| Polypeptone peptone | 10 g |
| Dextrose | 10 g |
| Sodium chloride | 5 g |
| L-cystine | 1 g |
| Agar | 15 g |

* Obtained from Traders Protein Division, Fort Worth, Texas.

Final pH is 6.8.

Suspend 25.4 g of the dehydrated medium in 500 ml of distilled water, mix thoroughly, and heat with frequent agitation until it boils for 1 minute. Autoclave 15 psi/15 minutes at 121° C.

Prepare 250 ml of sterile 2% hemoglobin solution and add to cystine heart agar. Dispense into screw-capped tubes and slant. Seal caps after cooling occurs.

### Hemoglobin Solution

Hemoglobin, 5 g, is added to 250 ml of water, shaken for 15 minutes, and autoclaved 15 psi/15 minutes at 121° C.

***Purpose***

For conversion of filamentous form to yeast from for *Histoplasma capsulatum* and *Sporothrix schenckii.*

## INHIBITORY MOLD AGAR (IMA) (BBL)

| | |
|---|---|
| Tryptone | 3 g |
| Beef extract | 2 g |
| Yeast extract | 5 g |
| Dextrose | 5 g |
| Starch (soluble) | 2 g |
| Dextrin | 1 g |
| Chloramphenicol | 0.125 g |
| Gentamicin | 5 mg |
| Salt A | 10 ml |
| Salt C | 20 ml |
| Agar | 17 g |
| Distilled water | 970 ml |

### Salt A

| | |
|---|---|
| $NaH_2PO_4$ | 25 g |
| $Na_2HPO_4$ | 25 g |
| $H_2O$ | 250 ml |

### Salt C

| | |
|---|---|
| $MgSO_4$-$7H_2O$ | 10 g |

| | |
|---|---|
| $FeSO_4$-$7H_2O$ | 0.5 g |
| NaCl | 0.5 g |
| $MnSO_4$-$7H_2O$ | 2.0 g |
| $H_2O$ | 250 ml |

Materials are dissolved in water which is brought to a boil to suspend the agar. After cooling, pH is adjusted to 6.7. Autoclave 15 psi/15 minutes. Chloramphenicol is first dissolved in 2 ml of alcohol (95%) and added to boiling medium. Pour into sterile, plastic Petri dishes (35 ml/plate).

***Purpose***

For isolation and subculture of fungi.

## L-DOPA FERRIC CITRATE REAGENT DISKS

I. Phosphate Buffer

| | |
|---|---|
| A. $Na_2HPO_4$ (0.067 M) | .951 g |
| Distilled water | 100.0 ml |
| B. $KH_2PO_4$ (0.067 M) | .912 g |
| Distilled water | 100.0 ml |

Mix equal volumes of A and B; pH to 6.8.

II. L-Dopa (L-B-3, 4-dihydroxy-phenylalanine) solution.
Suspend in 1–3 drops of dimethyl sulfoxide and dissolve (vigorous vortexing is required) in distilled water at a final concentration of 3 mg/ml (0.003 g/ml).

III. Ferric Citrate Solution
1 mg/ml (0.001 g/ml) in distilled water. Heat gently to dissolve.

IV. L-Dopa Ferric Citrate Solution

| | |
|---|---|
| L-Dopa solution | 3.0 ml |
| Ferric citrate solution | 1.5 ml |
| Phosphate buffer | 10.5 ml |

Final solution should be light blue in color.

V. Sterile blank paper disks (Difco) are each saturated with the L-Dopa ferric citrate solution. Disks should be dried in sterile petri dishes in 37° C incubator. Store disks in freezer at −20° C for up to 6 months.

***Purpose***

For rapid detection of phenoloxidase production by *Cryptococcus neoformans.*

## LACTRIMEL MEDIUM

| | |
|---|---|
| Whole wheat flour | 20 g |
| Skim milk (20 g/200 ml distilled water) | 200 ml |
| Honey | 10 g |
| Distilled water | 1000 ml |
| Agar | 20 g |

Autoclave (15 psi/15 min) milk separately, add ingredients together before dispensing into Petri dishes.

***Purpose***

For inducing sporulation of dematiaceous fungi.

## MYCOBIOTIC (MYCOSEL) AGAR (DIFCO) (BBL)

| | |
|---|---|
| Phytone peptone | 10.00 g/liter |
| Dextrose | 10.00 g |
| Agar | 15.00 g |
| Cycloheximide | 0.40 g |
| Chloramphenicol | 0.05 g |

Suspend 36 g of the dehydrated medium in a liter of distilled water. Mix thoroughly, and bring to a boil. Autoclave 12 psi/15 minutes at 118° C. Dispense 35–40 ml into Petri dishes. Final pH is 6.5 ± 0.2.

***Purpose***

For recovery of dermatophytes.

## NIGER SEED (BIRDSEED AGAR)

| | |
|---|---|
| Pulverized *Guizotia abyssinicia* seed* | 50 g |
| Agar | 15 g |

Add seed to 100 ml of distilled water and grind in a Waring Blendor. Boil for ½ hour in a liter of water. Strain through cloth to remove the water extract from the seed. Adjust volume of seed extract to 1000 ml with distilled water.

Add agar and boil until dissolved. Place in a flask and autoclave 15 psi/15 minutes. Final pH is 5.5. Dispense into sterile, plastic Petri dishes (35 ml/plate).

* Source: The Philadelphia Seed Company, P.O. Box 230, Plymouth Meeting, Pennsylvania 19462.

***Purpose***

For identification of *Cryptococcus neoformans.*

## NITRATE REDUCTION MEDIUM

| | |
|---|---|
| Potassium nitrate | 5.0 g |
| Sodium phosphate, monobasic | 11.7 g |
| Sodium phosphate, dibasic | 1.14 g |
| Zepharin chloride (1.2 ml of 17% solution) | 200.0 mg |
| Distilled water | 200.0 ml |

Standard, medium sized tipped, cotton swabs are saturated in the solution. Swabs are frozen, lyophilized, and autoclaved for 15 min at 121° C. An alternative is to dry the swabs by vacuum for 24 hours and autoclave. Swabs are stored in sterile containers.

For the rapid detection of nitrate reduction by cryptococci.

## POTATO DEXTROSE AGAR (BBL)

| | |
|---|---|
| Potato infusion | 200 g |
| Dextrose | 20 g |
| Agar | 15 g |
| Distilled water | 1000 ml |

Boil to put into solution. Adjust pH to 5.6. Autoclave 15 psi/15 minutes. Place sufficient medium in screwcapped tubes to give a proper slant or dispense into sterile, plastic Petri dishes (15 ml/plate).

***Purpose***

For pigment production by *Trichophyton rubrum.*

## RICE GRAIN MEDIUM

| | |
|---|---|
| Polished white rice (without added vitamins) | 8 g |
| Distilled water | 25 ml |

Autoclave in a 125-ml cotton-plugged Erlenmeyer flask at 15 psi/15 minutes. Inoculate with fragments of a growing culture.

***Purpose***

For identification of *Microsporum audouinii.*

### SABHI AGAR

| | |
|---|---|
| Calf brain infusion | 100 g |
| Beef heart infusion | 125 g |
| Proteose peptone | 5 g |
| Glucose | 21 g |
| Sodium chloride | 2.5 g |
| Disodium phosphate | 1.25 g |
| Agar | 15 g |

To rehydrate the medium, suspend 59 g in 1000 ml of distilled water and heat to boiling. Sterilize in autoclave, cool to 50° C and add 1 ml of sterile 100 mg/ml chloramphenicol solution. Dispense into tubes or plates.

***Purpose***

Useful for the recovery of *Blastomyces dermatitidis* and *Histoplasma capsulatum.* The addition of 10% sheep blood increases the recovery of *Histoplasma capsulatum.* Also useful for conversion of these two organisms into their yeast form.

### SABOURAUD'S DEXTROSE AGAR (EMMON'S MODIFICATION)

#### Broth

| | |
|---|---|
| Dextrose | 20 g |
| Neopeptone | 10 g |
| Distilled water | 1000 ml |

Dissolve by boiling and tube in 10-ml aliquots into tubes (8 × 150 mm). Autoclave at 15 psi/15 minutes. Final pH is 6.8 to 7.0.

#### Agar

Add 17 g of agar to the above formulation. Dispense in sterile plastic Petri dishes (35 ml/plate).

***Purpose***

For recovery, subculture, and identification of filamentous fungi.

## TRICHOPHYTON AGARS (DIFCO)

| | |
|---|---|
| Bacto-Trichophyton Agar 1 | |
| Bacto-vitamin free casamino acids | 2.5 g |
| Bacto-dextrose | 40.0 g |
| Monopotassium phosphate | 1.8 g |
| Magnesium sulfate | 0.1 g |
| Bacto-agar | 15.0 g |
| Bacto-Trichophyton Agar 2 | |
| Bacto-Trichophyton agar 1 with inositol | 50.0 mg |
| Bacto-Trichophyton Agar 3 | |
| Bacto-Trichophyton agar with inositol | 50.0 mg |
| Thiamine | 200.0 mg |
| Bacto-Trichophyton Agar 4 | |
| Bacto-Trichophyton agar 1 with thiamine | 200.0 mg |
| Bacto-Trichophyton Agar 5 | |
| Bacto-Trichophyton agar 1 with nicotinic acid | 2.0 g |
| Bacto-Trichophyton Agar 6 | |
| Bacto-Trichophyton agar 1 with ammonium nitrate | 1.5 g |
| Bacto-Trichophyton Agar 7 | |
| Bacto-Trichophyton agar 1 with ammonium nitrate and histidine | 30.0 g |

Each medium is prepared according to manufacturer's instructions. Each medium is autoclaved for 10 minutes at 120° C and is dispensed into screw-capped tubes.

### *Purpose*

For identification of members of the genus *Trichophyton.*

## TYROSINE AGAR

| | |
|---|---|
| Nutrient agar | 23 g/liter |
| Tyrosine | 5 g |
| Distilled water | 1 liter |

Dissolve tyrosine in 100 ml of distilled water. Dissolve nutrient agar in 900 ml of distilled water and combine with tyrosine solution. Adjust pH to 7.0 and autoclave at 15 psi/15 minutes at 121° C. Dispense 20 ml/plate and swirl medium so that crystals are evenly distributed.

***Purpose***

For identification of *Nocardia* and *Streptomyces.*

## UREA R BROTH (DIFCO)

### (Rapid Urease Test) (73)

| | |
|---|---|
| Bacto yeast extract | 0.1 g |
| Monopotassium phosphate | 0.091 g |
| Disodium phosphate | 0.095 g |
| Urea | 20.0 g |
| Phenol red | 0.01 g |

Urea R broth is supplied ready to use by the manufacturer. To use, add 3.0 ml distilled water to each vial. Reagent should be used immediately or on same day prepared.

***Purpose***

For rapid detection of urease production by cryptococci.

## UREA BASE (BBL)

### (Rapid Selective Urease Test) (92)

| | |
|---|---|
| Gelysate peptone | 1.0 g |
| Dextrose | 1.0 g |
| Soldium chloride | 5.0 g |
| Monopotassium phosphate | 2.0 g |
| Urea | 20.0 g |
| Phenol red | 0.012 g |

For preparation of standard urea base, 29 g of powder is dissolved in 100 ml distilled water and sterilized by filtration. For the rapid selective urease test, 29 g of powder is dissolved in 20 ml of water, effecting a 5× concentration. The pH is adjusted to 5.5 ± 0.05 and the reagent filter sterilized into a large test tube.

Sterile 6-inch long applicator sticks with small cotton tips are immersed in the concentrated urea base for 20–30 minutes. The sticks are removed, excess fluid from the saturated cotton tip is expressed onto sterile gauze and the sticks are placed into a tube suitable for lyophilization. The swabs are frozen for 1 hour at −70° C and lyophilized overnight.

1% benzalkonium chloride is prepared in sterile water. pH must be adjusted to 4.86 ± 0.01 (pH is critical).

***Purpose***

For the selective identification of *Cryptococcus neoformans.*

## YEAST EXTRACT AGAR

| | |
|---|---|
| Yeast extract | 1 g |
| Buffer* | 2 ml |
| Agar | 20 g |
| Distilled water | 1000 ml |

Boil into solution, sterilize 15 psi/15 minutes. Pour into sterile, plastic Petri dishes (35 ml/plate).

***Purpose***

For identification of *Histoplasma capsulatum, Blastomyces dermatitidis* and *Coccidioides immitis.*

## YEAST FERMENTATION BROTH

### Plain Broth

| | |
|---|---|
| Distilled water | 1000 ml |
| Peptone | 10 g |
| NaCl | 5 g |
| Beef extract | 3 g |
| 1 N NaOH | 1 ml |

Boil to dissolve.

**Indicator (Bromcresol Purple)**

Dissolve 0.04 g of bromcresol purple in 100 ml of distilled water. Add a small amount of 1 N NaOH to make an alkaline solution. Let stand overnight.

After the dye is in solution, add 1 N HCl until the neutral point is reached and 1 drop of either acid or base will cause a complete color change.

Add 100 ml of indicator to 1 liter of plain broth. Tube (9 ml/tube) in 18- × 150-mm tubes with inverted Durham tubes. Sterilize 15 psi/15 minutes.

---

* Dissolve 40 g of $Na_2HPO_4$ in 300 ml of distilled $H_2O$; then add 60 g of $KH_2PO_4$. The pH is 6.0. If necessary, adjust with 1 N HCl or NaOH. Adjust the volume to 400 ml with distilled $H_2O$ and store at 4° C.

## YEAST NITROGEN BASE AGAR (DIFCO)

Prepare a 2% agar solution (20 g of agar/liter of distilled water). Autoclave at 15 psi/15 minutes at 121° C.
Prepare yeast nitrogen base (10× concentration) by dissolving 6.7 g of base/100 ml of distilled water. pH from 6.2 to 6.4 or to the point of precipitation by adding 1 N NaOH. Discard 12 ml of solution. Sterilize by filtration using a Nalgene filter.

Add 88 ml of yeast nitrogen base and 100 ml of filter-sterilized bromcresol purple indicator to 1 liter of 2% agar solution. Pour into sterile, plastic Petri dishes (20 ml/plate).

***Purpose***

For identification of yeasts based on carbohydrate utilization.

# TEST SOLUTIONS, STAINS AND REAGENTS

## BROMCRESOL PURPLE INDICATOR

Dissolve 0.04 g of bromcresol purple in 100 ml of distilled water. Add a small amount of 1 N NaOH to make alkaline. Allow to stand overnight.
After the dye is in solution, add 1 N HCl until the neutrality is reached and 1 drop of either acid or base will cause a complete change in color.

***Purpose***

For use in carbohydrate utilization and fermentation media.

## CALCOFLUOR WHITE STAIN (29a)

Prepare a 1% (wt/vol) stock solution of calcofluor white M2R (power available from Polysciences, Inc, Warrington, PA 18976) by dissolving 1 g of the powder in 100 ml distilled water with gentle heating.
Working solution of KOH (for staining of skin scales, hair, nail scrapings and "dirty" body fluids): 10% (wt/vol) potassium hydroxide (KOH).
Working solution of calcofluor white (for staining of tissue sections): 0.1% calcofluor white containing 0.05% Evans blue for counter stain. Add 1 drop of KOH and 1 drop of calcofluor white to specimen.

***Purpose***

For sharper delineation of fungal elements in clinical specimens by fluorescence microscopy.

## CARBOHYDRATE SOLUTIONS

Add 20 g of carbohydrate to 100 ml of distilled water. Dissolve by placing in a 56° C water bath for a few minutes. Sterilize by filtration through Nalgene filter.

Add 0.5 ml of stock carbohydrate solutions to the cooled tubes of plain broth and indicator just before use.

### *Purpose*

For identification of yeasts.

## GASTRIC MUCIN

Emulsify 5 g of granular gastric mucin in 100 ml of distilled water in a Waring Blendor for 5 minutes. Autoclave 15 psi/15 minutes. Cool to room temperature. Adjust the pH to 7.3. Check the sterility by inoculating one blood agar plate. Store at 4° C.

### *Purpose*

Useful for inducing in vivo conversion of dimorphic pathogens by animal inoculation.

## INDIA INK OR NIGROSIN STAINING SOLUTION

| | |
|---|---|
| Nigrosin (granular)* | 10 g |
| Formalin (10%)† | 100 ml |

Place the above solution in a boiling water bath for 30 minutes. Add 10% formalin lost by evaporation. Filter twice through double filter paper (Whatman No. 1). India ink (Pelican Brand) may be used after the addition of 40% formalin to make a final concentration of 10%.

### *Purpose*

For detection of cryptococci in cerebrospinal fluid.

## LACTOPHENOL-ANILINE BLUE STAIN

| | |
|---|---|
| Distilled water | 20 ml |
| Lactic acid | 20 ml |
| Phenol crystals | 20 g |

* Available commercially from Harleco (Division of Hartman-Ledden Co.), Philadelphia, Pennsylvania.

† 1:10,000 merthiolate dilution is acceptable also.

| | |
|---|---|
| Aniline blue (Cotton blue; Poirrier's blue) | 0.05 g |
| Glycerol | 40 ml |

Dissolve phenol in the lactic acid, glycerol, and water by gently heating. Then add aniline blue.

***Purpose***

For wet mount preparations of cultures when a small amount of agar is removed with the culture.

## LACTOPHENOL-ANILINE BLUE POLYVINYL ALCOHOL STAIN

| | |
|---|---|
| Polyvinyl alcohol | 15 g |
| Distilled water | 100 ml |
| Lactic aid | 39 ml |
| Phenol (melted) | 39 ml |
| Aniline blue (Poirrier's blue) | 0.1 g |

Add polyvinyl alcohol to the water. Place in an 80° C water bath. Stir until smooth and the solution clears. Add the lactic acid and the phenol. **Always add lactic acid before phenol.**

Add 0.1 g of aniline blue (0.05%).

***Purpose***

For wet mount or scotch tape preparations of cultures. The polyvinyl alcohol acts to minimize evaporation so that mounts can be retained longer for study, however, it reacts with agar and cannot be used when a small amount of agar is removed when making a wet mount.

## NITRATE REDUCTION TEST REAGENTS

| | |
|---|---|
| Reagent A | |
| Glacial acetic acid | 50 ml |
| Water | 125 ml |
| Sulfanilic acid | 1.4 g |
| Reagent B | |
| Glacial acetic acid | 50 ml |
| Water | 125 ml |
| Dimethyl $\alpha$-naphthylamine | 1 g |

It may take overnight for reagents to go into solution. Frequent agitation helps.

***Purpose***

Helpful in the identification of yeasts.

## KINYOUN'S ACID-FAST STAIN (MODIFIED)

### Carbolfuchsin Solution

| | |
|---|---|
| Basic fuchsin | 4.0 g |
| 95% ethyl alcohol | 20.0 ml |

Dissolve dye in alcohol and allow to stand for 24 hours. Add:

| | |
|---|---|
| Phenol (conc.) | 8.0 ml |
| Distilled water | 100.0 ml |

### Destaining Agent: 0.5% Aqueous $H_2SO_4$

| | |
|---|---|
| Sulfuric aid (conc.) | 0.5 ml |
| Distilled water | 99.5 ml |

### Methylene Blue Solution

| | |
|---|---|
| Methylene blue | 0.3 g |
| 95% ethyl alcohol | 30.0 ml |

Dilute to 100 ml with distilled water.

### Staining Procedure

Apply to heat-fixed smears*:

1. Kinyoun's carbol fuchsin. Flood the slide with stain for 5 minutes at room temperature. Do not heat.
2. Rinse with water.
3. Rinse with 50% ethyl alcohol. Flood and pour off until excess red dye is removed.
4. Water rinse.
5. Rinse with 0.5% aqueous $H_2SO_4$ for approximately 3 minutes.
6. Counterstain with methylene blue for 1 minute.
7. Rinse with water. Blot dry and examine under oil immersion.

* Control smears of *Nocardia asteroides* (filaments partially acid-fast) and *Streptomyces* species (nonacid-fast) should be run with each group of unknowns.

***Purpose***

To stain clinical materials in suspected cases of nocardiosis; or, to stain cultures suspected of being *Nocardia* species.

## KOH (POTASSIUM HYDROXIDE)

| | |
|---|---|
| Potassium hydroxide (crystals) | 10 g |
| Glycerin | 10 ml |
| Distilled water | 80 ml |

***Purpose***

For detection of fungal elements in clinical specimens by microscopic examination.

# COLOR PLATES

# COLOR PLATE 1

## THE DEMATIACEOUS MOLDS

Illustrated in the color plate on the opposite page are several colonial variants produced by different species of dematiaceous molds.

**Figure 1** illustrates the deep brown to black color produced by this group of fungi. The reverse of the colony is equally as black. This colony is characteristic of *Alternaria* sp or *Curvularia* sp.

**Figure 2** is also representative of *Alternaria*, *Curvularia*, or *Helminthosporium.* Note that the entire central portion of the colony has been overgrown with a cottony, sterile mycelium, leaving only a thin concentric black ring at the periphery where sporulation is still active.

**Figure 3** also illustrates the overgrowth of a very black almost yeast-like colony with a sterile, white mycelium. This particular colony is of *Stemphylium* sp.

**Figure 4** illustrates the variegated play of colors: red, orange and yellow against black, characteristic of *Epicoccum* sp.

**Figure 5, 6,** and **7** are colonial variants of the *Cladosporium*, *Phialophora*, and *Fonsecaea* sp of dematiaceous molds.

**Figure 5** is more characteristic of the rapidly growing, saprobic species of *Cladosporium.* These saprobes often produce a dark green-black variant. **Figure 6** is a leathery, rugose variant, while **Figure 7** is the variant having the velvety or hair-like texture due to the production of a low aerial mycelium. **Figures 6** and **7** are more commonly seen with the slow growing, pathogenic species associated with chromoblastomycosis or mycetoma.

**Figure 8** reveals a glabrous, yeasty colony. This type of black yeast is characteristic of *Aureobasidium pullulans*; however, it must be remembered that the *Exophiala or Wangiella* group of molds may develop early as a black yeast similar to *Aureobasidium*; however, with a few days of additional incubation, a characteristic mycelium as seen in **Figure 7** begins to develop and the true nature of the "yeast" can be determined.

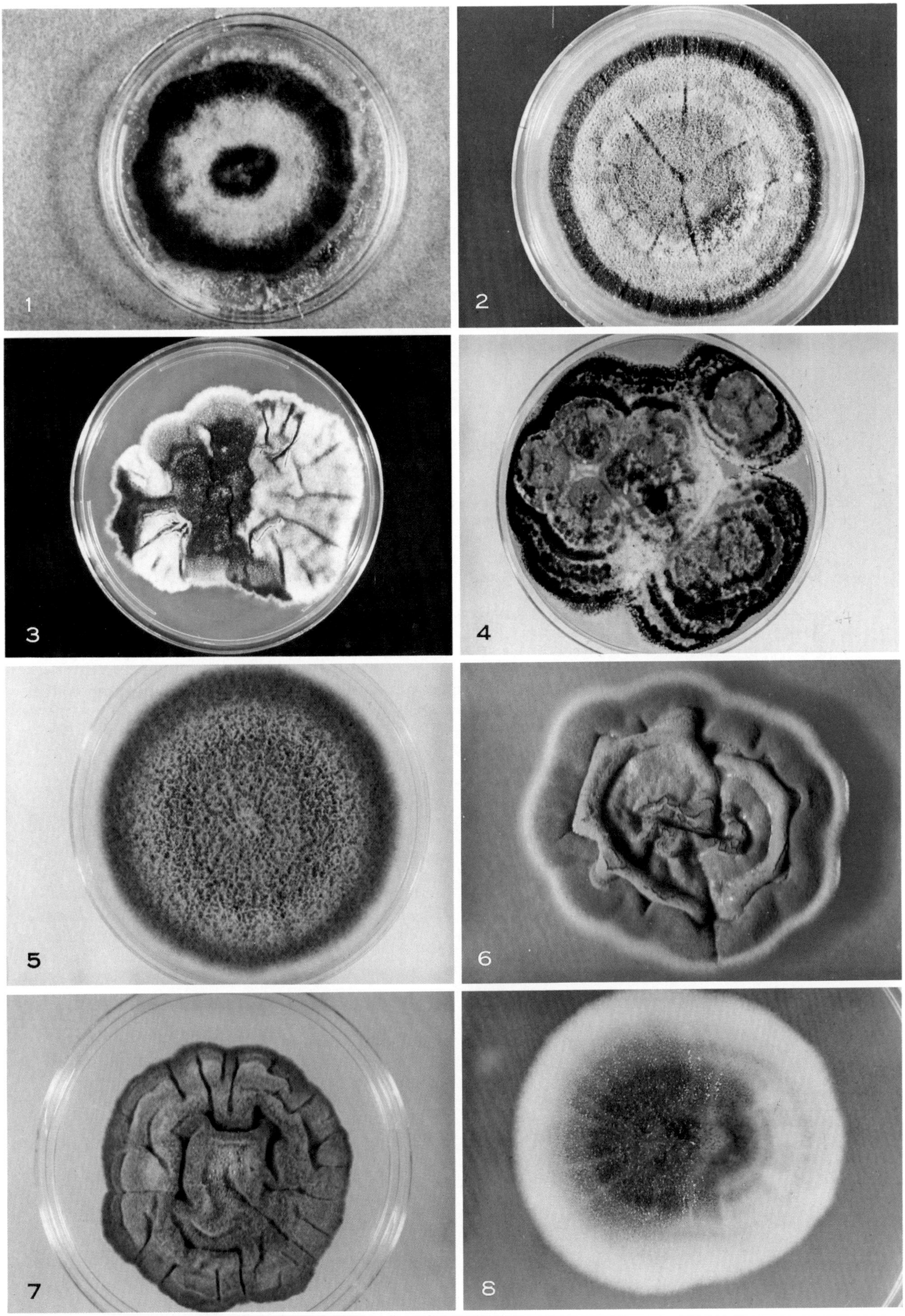
1
2
3
4
5
6
7
8

# COLOR PLATE 2

## THE HYALINE RAPIDLY GROWING MOLDS

On the opposite page are illustrated several colonies produced by members of the rapidly growing hyaline molds.

**Figure 1** illustrates three separate agar plates, one containing 20% sucrose agar (*upper left*), the second maltose agar (*upper right*), and the third, Czapek's agar (*lower center*). These three agars were multiply inoculated with the same strain of *Aspergillus glaucus* and incubated at 25° C for the same length of time. Note the wide variation in colonial morphology with the different types of media used. When describing gross colonial morphology, the type of medium and the environmental conditions for culture must always be specified. The remaining colonies in this color plate were grown on Sabouraud's dextrose agar and incubated at room temperature.

**Figure 2** is a typical colony of *Penicillium* sp showing the green surface and radial rugae characteristic of this fungus. Strains of *A. fumigatus* may appear similar, and microscopic examination is always necessary before a definitive identification can be made.

**Figure 3** is highly suspicious for *A. flavus* on visual inspection; however, microscopic study revealed the typical fruiting bodies of *Penicillium* sp. This somewhat surprising colonial morphology again illustrates the need for microscopic study.

**Figure 4** is a variant of *Paecilomyces* sp, but this type of colony could also be produced by *Scopulariopsis* sp. *Paecilomyces* sp also include strains that produce light green or grey-green colonies closely simulating *Penicillium* sp.

**Figure 5** is a classic colony of *Scopulariopsis* sp. The granular, buff to brown, rugose surface appearance is classic. *Scopulariopsis* sp are always some shade of buff or brown and never produce any of the green-blue or green-yellow pastels.

**Figure 6** is a fairly characteristic colony of *Acremonium* (*Cephalosporium*) sp. The colony often has this cream or light peach color and is smooth and almost yeast-like due to the production of a delicate low aerial mycelium. *Acremonium* (*Cephalosporium*) sp may be confused both in gross colonial appearance and in microscopic morphology with *Sporothrix schenckii* mold form which never produces a fluffy aerial mycelium.

**Figure 7** is the characteristic green lawn of either *Gliocladium* or *Trichoderma* sp. Yellow or yellow-green variants may also be encountered.

**Figure 8** illustrates the orange-red color characteristics of *Fusarium* sp. The pigment produced is water soluble and tends to discolor the agar. Purple, lavender, and violet variants are also commonly encountered.

1
20%
CZ
M
2
3
4
5
6
7
8

# COLOR PLATE 3

## THE DERMATOPHYTES

On the opposite page are illustrated several colonial variants produced by the dermatophytic molds. Because the colonial morphology is not consistent and considerable overlap occurs between members of this group, microscopic examination is always required before a genus or species identification can be confirmed.

**Figure 1** is a colony of *Microsporum audouinii*, revealing a velvety to fluffy, tan mycelium. A central nidus of sterile, white fluffy mycelial growth is seen. The salmon-brown pigmentation noted off center to the left is characteristic of *M. audouinii*, a coloration also seen on the reverse of the colony.

**Figure 2** is a colony of *M. canis*, revealing a cottony, cream to buff aerial mycelium. The lighter lemon yellow to yellow-tan apron at the periphery is characteristic for *M. canis* and this species should always be suspected when this colonial type is observed.

**Figure 3** is a colony of *M. gyspeum*, which usually sporulates very heavily and produces a granular or sugary colony. The color ranges from white to buff-yellow, and the starburst appearance seen in the colony shown is suggestive of *M. gypseum*.

**Figure 4** is a culture plate on which are growing several colonies of *Epidermophyton floccosum*. Colonies of *E. floccosum* are generally white to light tan and have a fluffy (floccose) surface. A yellow-tan or buff outer apron is not unusual, reminiscent of *M. canis*.

**Figure 5** reveals an entire colony, velvety in appearance, covered with a low aerial mycelium, with irregular radial rugae. This colonial type may be seen with a number of dermatophytes, notably *Trichophyton tonsurans*, *T. verrucosum*, and *T. schoenleinii*. The different growth rates, ability to grow on the several *Trichophyton* agars, and characteristic microscopic appearance must be assessed before identification is possible.

**Figure 6** is a characteristic colonial variant of *T. mentagrophytes* or *T. rubrum*, granular type. Fluffy variants also may be encountered. Note the umbonate center of sterile growth. These sterile tufts should be avoided when subculture or microscopic study of the colony is desired; rather, the granular, sporulating areas should be selected.

**Figure 7** illustrates the reverse of a colony of *T. rubrum*, revealing the deep wine red pigment that is characteristically formed by this species. However, red or orange pigments may also be produced by other dermatophytes, notably *T. mentagrophytes* and *T. ajelloi*, and is not diagnostic of *T. rubrum*.

**Figure 8** illustrates the smooth, waxy, lavender to purple appearance of *T. violaceum*. These colonies are extremely slow growing. The waxy appearance results from the inability of this species to produce an aerial mycelium.

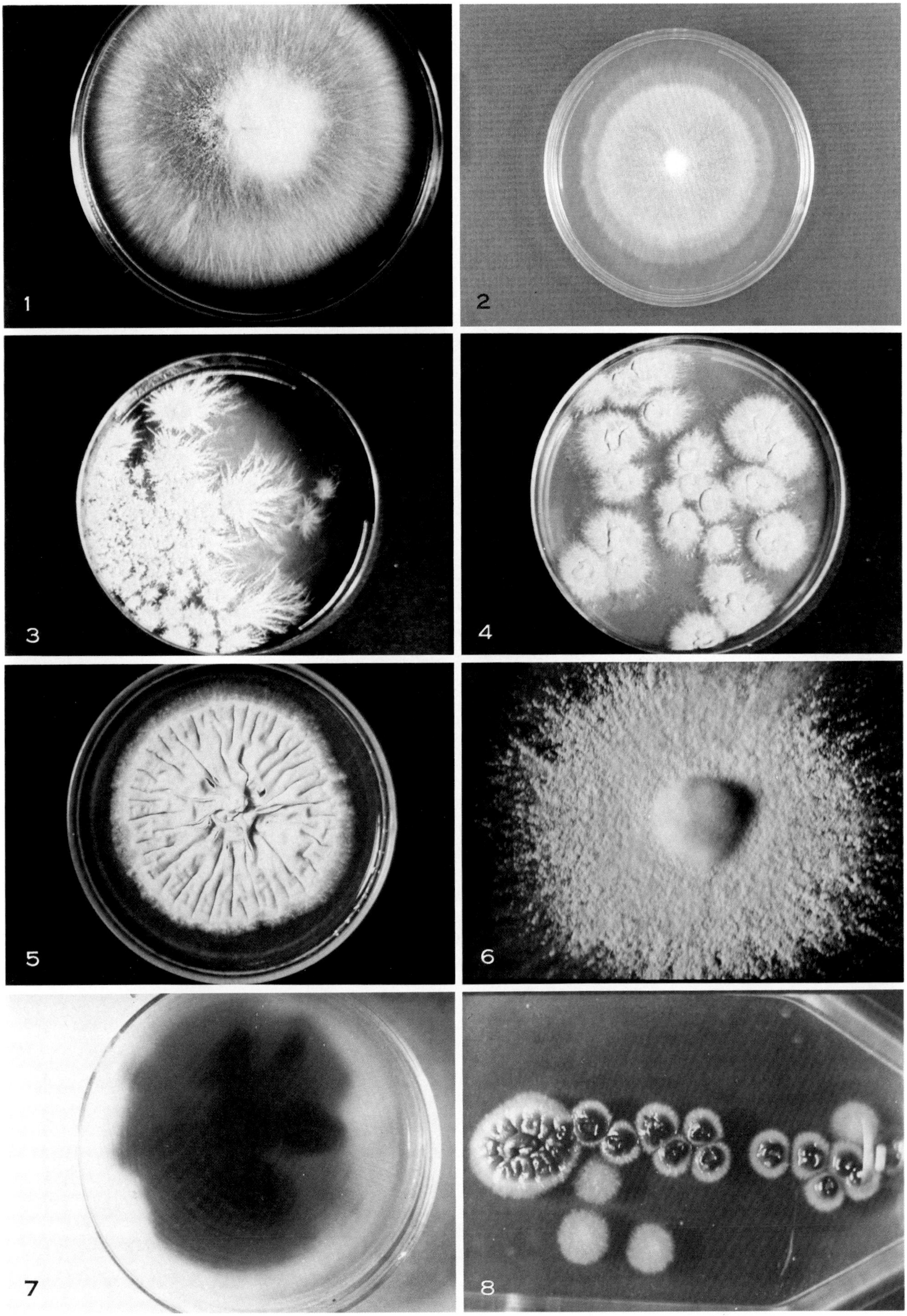
1
2
3
4
5
6
7
8

# COLOR PLATE 4

## THE OPPORTUNISTIC PATHOGENIC FUNGI

On the opposite page are a number of colonies produced by fungi or fungus-like bacteria that may cause opportunistic infections in man, particularly in debilitated hosts.

**Figure 1** is a characteristic colony of *Aspergillus fumigatus*, revealing the dark green granular center of the colony, surrounded by a white apron of sterile growth.

**Figure 2** is a colony of *A. flavus.* A number of other hyaline molds may have similar appearing colonies, including other species of *Aspergillus.* Microscopic examination is required to make a final identification.

**Figure 3** is a characteristic colony of *A. niger.* The surface black appearance of the colony is due to the dense production of jet black spores. Although this appearance may suggest one of the dematiaceous molds, the reverse of the colony is light tan or buff, never black.

**Figures 4** and **5** are young and old colonies of one of the Zygomycetes (Phycomycetes). These colonies grow extremely rapidly, covering the entire agar surface with a fluffy mycelium. The mycelium is initially white as seen in **Figure 4;** however, with maturity and the production of spores, a dark gray or brown discoloration may be observed.

**Figure 6** illustrates the yellow, wrinkled, smooth, brittle colonies of *Nocardia asteroides*.* Microscopic examination is required to detect the delicate branching filaments, in that yeast colonies can look quite similar. A musty basement odor is helpful in ruling out yeast colonies.

**Figure 7** illustrates the chalky white, brittle colonies of *Streptomyces* sp.* Again, the detection of a musty odor is helpful in the preliminary identification of these microorganisms.

**Figure 8** is the classic molar tooth colony of *Actinomyces israelii*.* as seen after 5 days incubation in an anaerobic environment on brain-heart infusion blood agar.

---

* Are classified as bacteria.

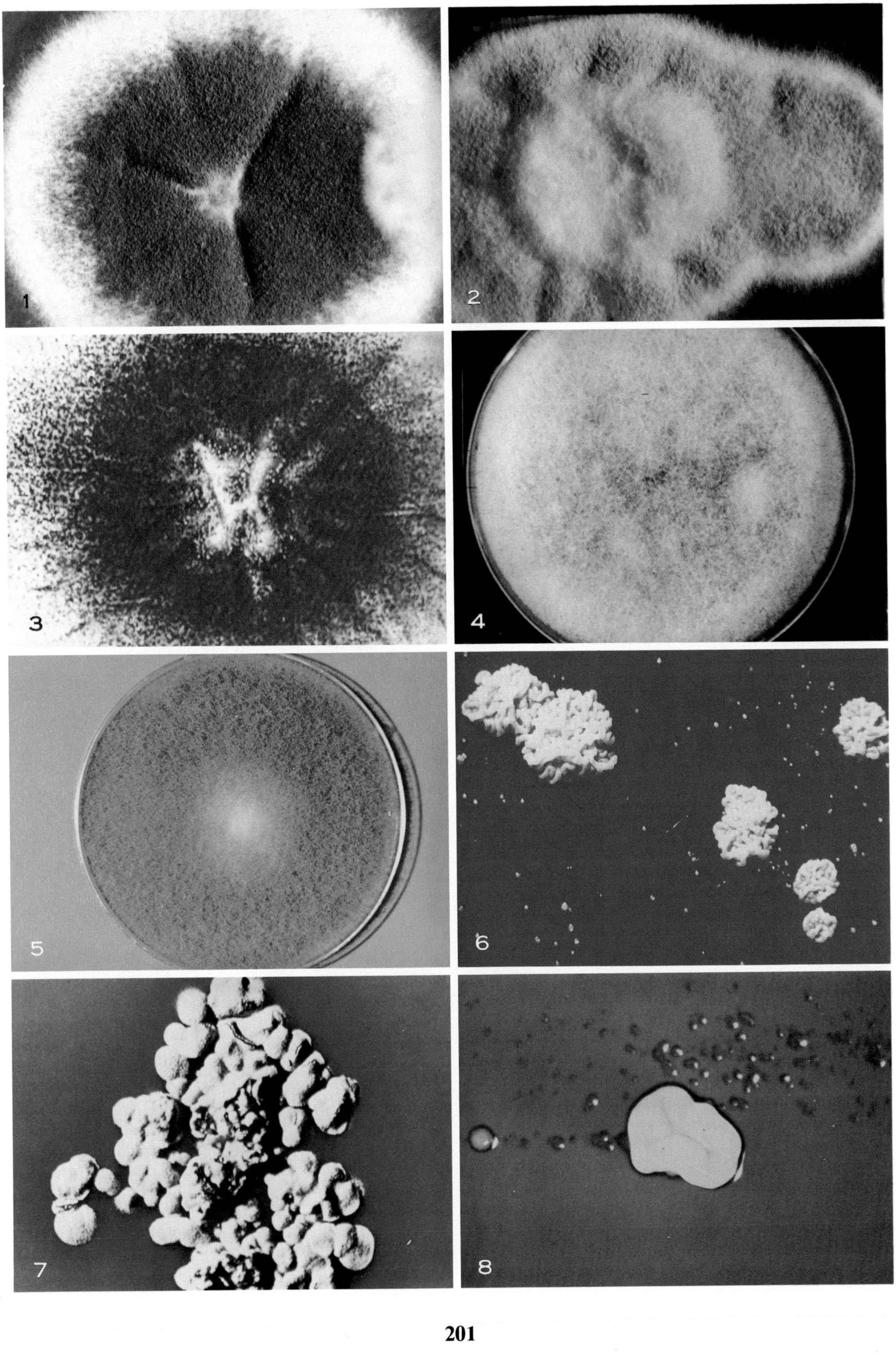
1
2
3
4
5
6
7
8

# COLOR PLATE 5

## THE DIMORPHIC PATHOGENS

On the opposite page are several colonies illustrating several of the dimorphic pathogens. Colonies of these types must always be handled carefully by laboratory personnel as the chance for self-infection is high. A properly operating biological safety hood should always be used when transferring these colonies for further study.

**Figure 1** illustrates the phenomenon of dimorphism, showing partial conversion to a yeast form in the upper left portion of the colony, with the lower right covered by the unconverted mycelial form of the fungus.

**Figure 2** is a characteristic silky smooth, white to brown colony characteristic of *Blastomyces dermatitidis* and *Histoplasma capsulatum*, mycelial form. The brownish discoloration often darkens as the colony ages.

**Figure 3** is the yeast form conversion characteristic of *Blastomyces dermatitidis* and of *H. capsulatum*, showing central coremia where conversion is not quite complete.

**Figure 4** is the fluffy mycelial form of *Coccidioides immitis*; however, this type of colony also could be produced by *Blastomyces dermatitidis.* A tube or plate showing this type of fluffy, almost cobweb-type of growth should never be opened outside of a biological safety hood, particularly if the growth is delayed beyond 5 days.

**Figure 5** is a yeast colony illustrating complete conversion of one of the dimorphic molds. This type of yeast colony may be seen most commonly with *B. dermatitidis* and *Paracoccidioides brasiliensis.*

**Figure 6** is also a photograph of the mycelial form of *C. immitis*, again showing the fluffy, cobweb nature of the colony. It is extremely difficult to convert *C. immitis* mold into the spherule form in the laboratory, and the diagnosis is usually made by demonstrating the alternately staining, barrel-shaped arthroconidia. Again, **take extreme care when studying this type of colony.**

**Figure 7** is a mold colony of *P. brasiliensis.* The colony is very slow growing and this photograph shows the propensity for this fungus to form heaped craters, with the agar dug out within the depths of the crater.

**Figure 8** illustrates the yeast form of *Sporothrix schenckii.* The mycelial form often also appears waxy or yeast-like in nature and it may be difficult to differentiate the two forms on gross colony examination alone. The mycelial form often turns dark brown or gray-black with age.

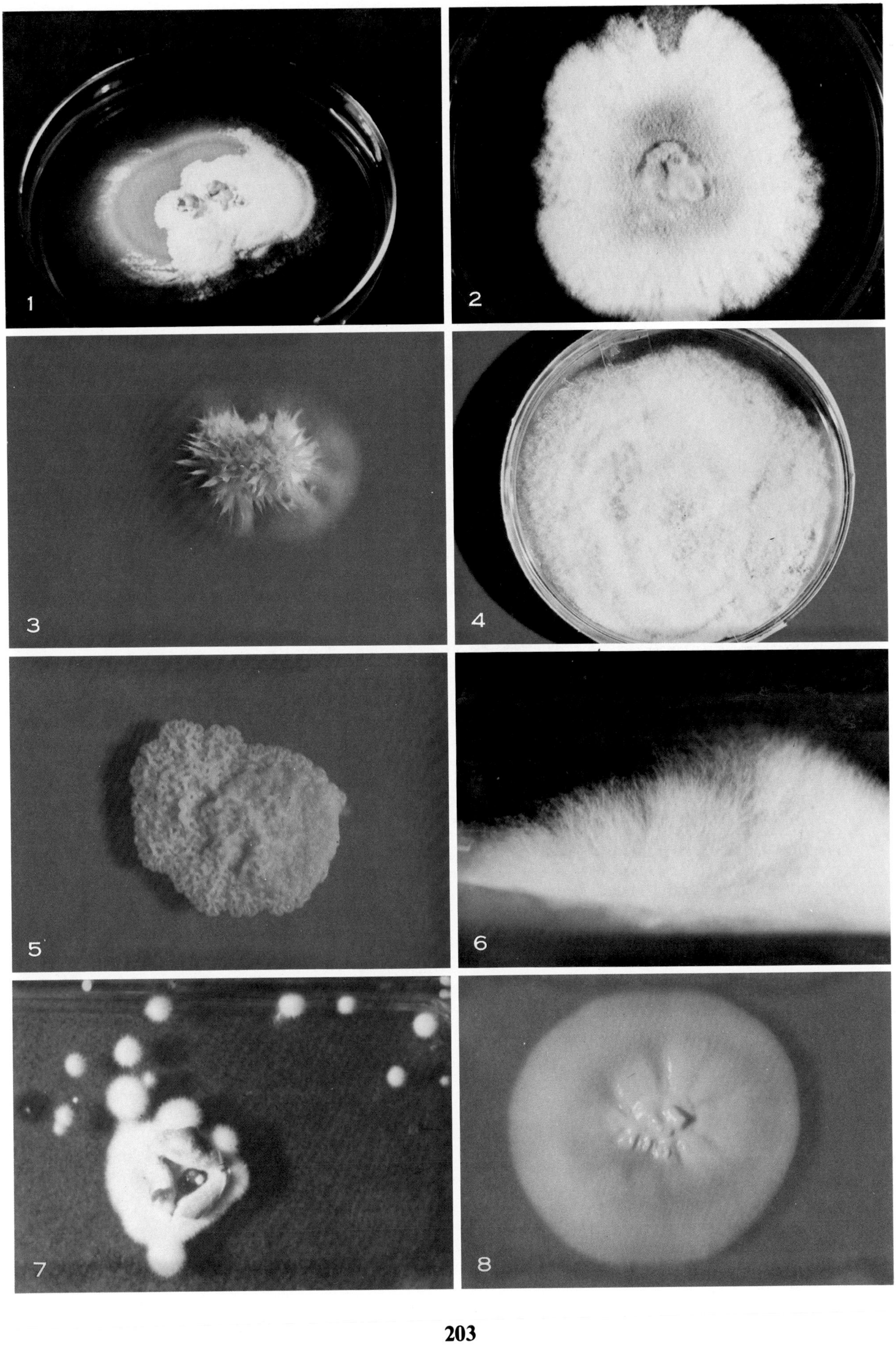
1
2
3
4
5
6
7
8

# COLOR PLATE 6

## YEAST IDENTIFICATION SYSTEMS AND TESTS

**Figure 1** illustrates uninoculated API 20C strip.

**Figure 2** illustrates an uninoculated Uni-Yeast-System.

**Figure 3** illustrates inoculated API Yeast-Ident System showing positive and negative reactions.

**Figure 4** is the C/N Screen tubes illustrating positive reactions (brown pigmentation, *left* two tubes) and negative controls (*right* two tubes).

**Figure 5** depicts the rapid selective urease test illustrating positive reaction (red) and negative control (colorless).

**Figure 6** is a niger seed agar plate illustrating brown pigmentation of colonies characteristic of *Cryptococcus neoformans*.

**Figure 7** is an auxanographic carbohydrate assimilation plate. Growth occurs around those filter paper disks that contain a carbohydrate that can be utilized by the yeast being tested.

**Figure 8** is a smear preparation of yeast cells from ascospore agar illustrating acid-fast staining ascospores (red staining cells) and blastoconidia (blue staining cells) characteristic of *Saccharomyces* sp.

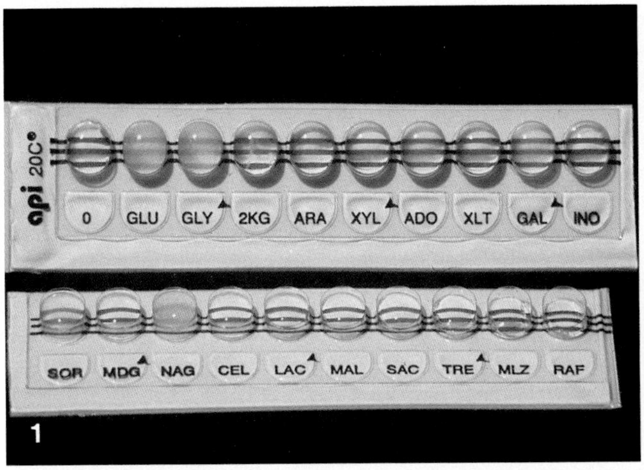
api 20C®
0 GLU GLY 2KG ARA XYL ADO XLT GAL INO
SOR MDG NAG CEL LAC MAL SAC TRE MLZ RAF
1

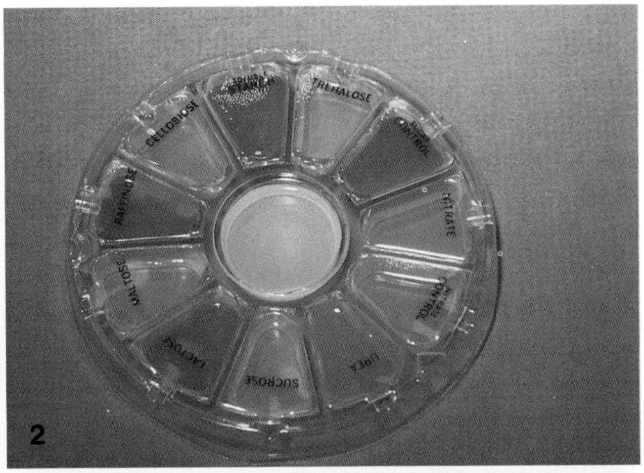
TREHALOSE
CELLOBIOSE
NITRATE
SUCROSE
UREA
LACTOSE
MALTOSE
2

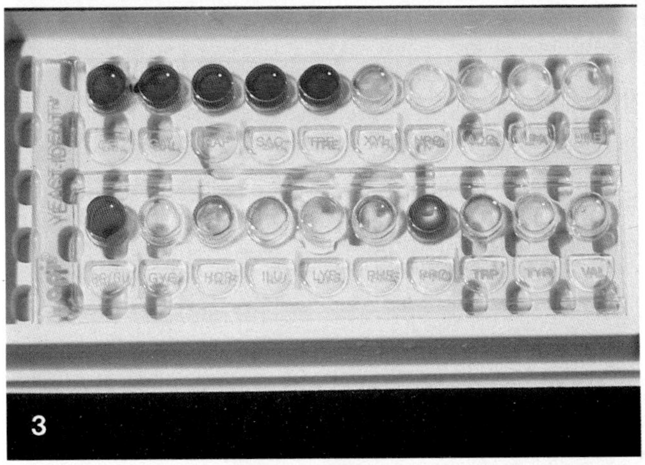
3

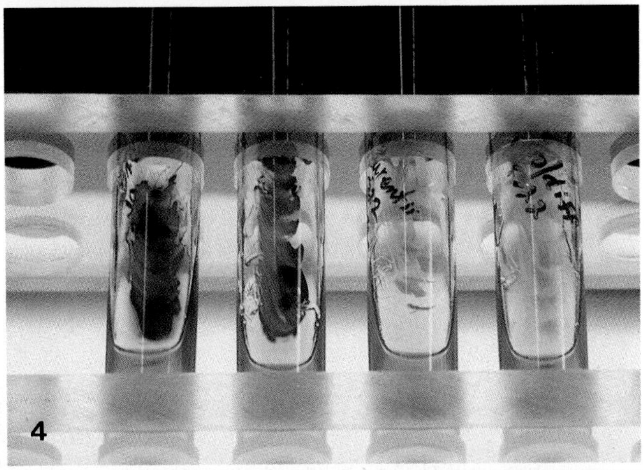
4

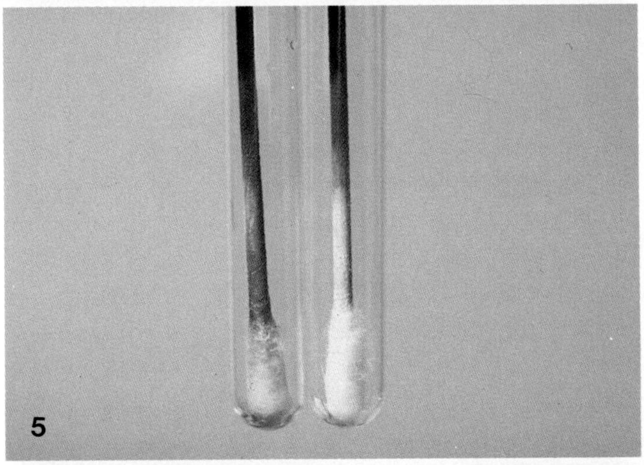
5

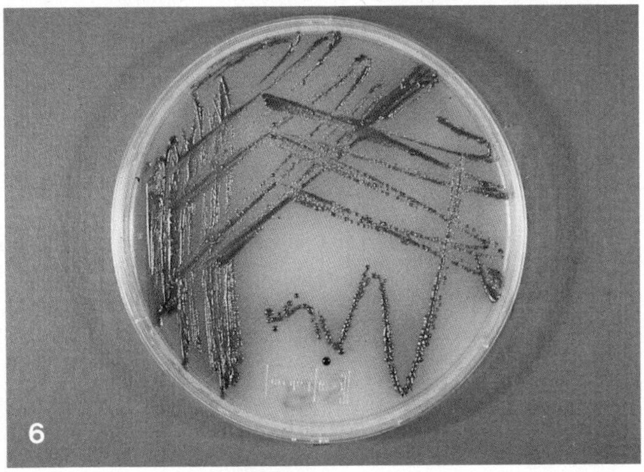
6

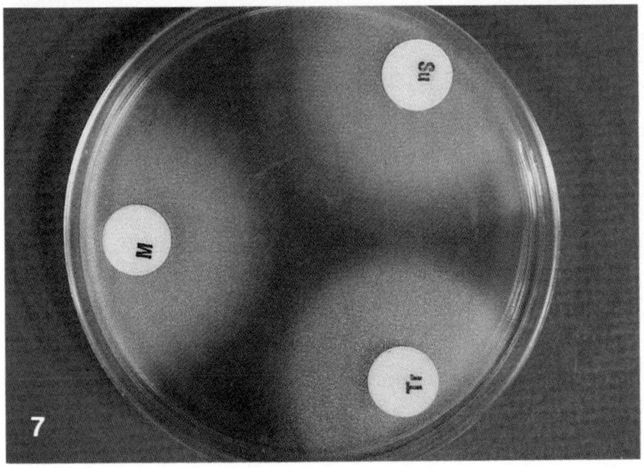
M
7

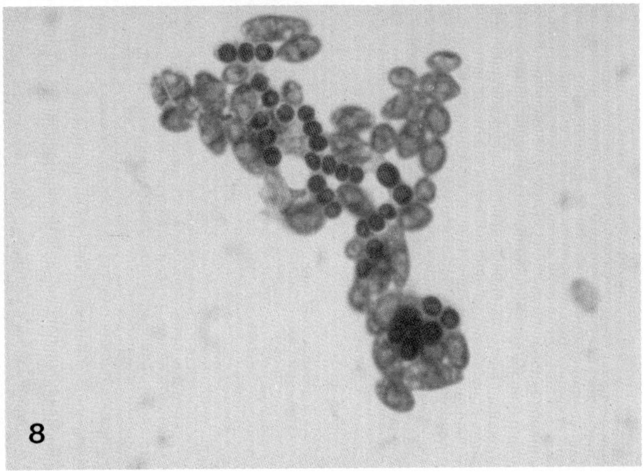
8

# INDEX